Lecture Notes in Physics

The Editorial Policy for Proceedings

The series Lecture Notes in Physics reports new developments in physical research and teaching – quickly, informally, and at a high level. The proceedings to be considered for publication in this series should be limited to only a few areas of research, and these should be closely related to each other. The contributions should be of a high standard and should avoid lengthy redraftings of papers already published or about to be published elsewhere. As a whole, the proceedings should aim for a balanced presentation of the theme of the conference including a description of the techniques used and enough motivation for a broad readership. It should not be assumed that the published proceedings must reflect the conference in its entirety. (A listing or abstracts of papers presented at the meeting but not included in the proceedings could be added as an appendix.)

When applying for publication in the series Lecture Notes in Physics the volume's editor(s) should submit sufficient material to enable the series editors and their referees to make a fairly accurate evaluation (e.g. a complete list of speakers and titles of papers to be presented and abstracts). If, based on this information, the proceedings are (tentatively) accepted, the volume's editor(s), whose name(s) will appear on the title pages, should select the papers suitable for publication and have them refereed (as for a journal) when appropriate. As a rule discussions will not be accepted. The series editors and Springer-Verlag will normally not interfere with the detailed editing except in fairly obvious cases or on technical matters.

Final acceptance is expressed by the series editor in charge, in consultation with Springer-Verlag only after receiving the complete manuscript. It might help to send a copy of the authors' manuscripts in advance to the editor in charge to discuss possible revisions with him. As a general rule, the series editor will confirm his tentative acceptance if the final manuscript corresponds to the original concept discussed, if the quality of the contribution meets the requirements of the series, and if the final size of the manuscript does not greatly exceed the number of pages originally agreed upon. The manuscript should be forwarded to Springer-Verlag shortly after the meeting. In cases of extreme delay (more than six months after the conference) the series editors will check once more the timeliness of the papers. Therefore, the volume's editor(s) should establish strict deadlines, or collect the articles during the conference and have them revised on the spot. If a delay is unavoidable, one should encourage the authors to update their contributions if appropriate. The editors of proceedings are strongly advised to inform contributors about these points at an early stage.

The final manuscript should contain a table of contents and an informative introduction accessible also to readers not particularly familiar with the topic of the conference. The contributions should be in English. The volume's editor(s) should check the contributions for the correct use of language. At Springer-Verlag only the prefaces will be checked by a copy-editor for language and style. Grave linguistic or technical shortcomings may lead to the rejection of contributions by the series editors. A conference report should not exceed a total of 500 pages. Keeping the size within this bound should be achieved by a stricter selection of articles and not by imposing an upper limit to the length of the individual papers. Editors receive jointly 30 complimentary copies of their book. They are entitled to purchase further copies of their book at a reduced rate. As a rule no reprints of individual contributions can be supplied. No royalty is paid on Lecture Notes in Physics volumes. Commitment to publish is made by letter of interest rather than by signing a formal contract. Springer-Verlag secures the copyright for each volume.

The Production Process

The books are hardbound, and the publisher will select quality paper appropriate to the needs of the author(s). Publication time is about ten weeks. More than twenty years of experience guarantee authors the best possible service. To reach the goal of rapid publication at a low price the technique of photographic reproduction from a camera-ready manuscript was chosen. This process shifts the main responsibility for the technical quality considerably from the publisher to the authors. We therefore urge all authors and editors of proceedings to observe very carefully the essentials for the preparation of camera-ready manuscripts, which we will supply on request. This applies especially to the quality of figures and halftones submitted for publication. In addition, it might be useful to look at some of the volumes already published. As a special service, we offer free of charge LATEX and TEX macro packages to format the text according to Springer-Verlag's quality requirements. We strongly recommend that you make use of this offer, since the result will be a book of considerably improved technical quality. To avoid mistakes and time-consuming correspondence during the production period the conference editors should request special instructions from the publisher well before the beginning of the conference. Manuscripts not meeting the technical standard of the series will have to be returned for improvement.

For further information please contact Springer-Verlag, Physics Editorial Department II, Tiergartenstrasse 17, D-69121 Heidelberg, FRG

E. Maruyama H. Watanabe (Eds.)

Physics and Industry

Proceedings of the Academic Session of the
XXI General Assembly of the International Union
of Pure and Applied Physics
Held at Nara, Japan, 22 and 23 September 1993

Springer-Verlag Berlin Heidelberg GmbH

Editors

Eiichi Maruyama
Angstrom Technology Partnership
Ryukakusan Building 8F
2-5-12 Higashi-Kanda
Chiyoda-ku Tokyo 101, Japan

Hisatsune Watanabe
NEC Corporation
Microelectronics Research Laboratories
1120 Shimokuzawa, Sagamihara
Kanagawa 229, Japan

Science Council of Japan
Physical Society of Japan
Japan Society of Applied Physics

ISBN 978-3-662-13966-0   ISBN 978-3-540-48676-3 (eBook)
DOI 10.1007/978-3-540-48676-3

CIP data applied for

Originally published by Springer-Verlag Berlin Heidelberg New York in 1994
Softcover reprint of the hardcover 1st edition 1994

Typesetting: Camera ready by author/editor
SPIN: 10127040   55/3140-543210 - Printed on acid-free paper

PREFACE

The XXI General Assembly of the International Union of Pure and Applied Physics (IUPAP) was held at the Nara New Public Hall in the City of Nara, Japan, during the period September 20 - 25, 1993. This was the first General Assembly ever to be held in Japan. It also marked the 71st anniversary of the Union.

The 1993 Assembly was held under the auspices of IUPAP and organized by the Science Council of Japan, the Physical Society of Japan, and the Japan Society of Applied Physics, with the cooperative support of the Association of Asia Pacific Physical Societies. The National Organizing Committee was chaired by Prof. Hiroshi Takuma of the University of Electro-Communications.

Representatives of 41 countries and one region were present at the General Assembly and elected Officers of the Union, Council Members, Commission Members and Delegates to Inter-Union Commissions for the next three years. They also discussed important issues related to pure and applied physics in the changing world.

The Academic Session with the titled "Physics and Industry" was held on September 22nd and 23rd. This volume includes all the presentations made at the Academic Session by seven speakers. The scope of the session is explained properly in the Introductory Remarks of Prof. Takuma. The organizers would like to express their heartfelt thanks to all the speakers who disclosed their stimulating views on the diverse problems concerning the relations between physics and industry. It is hoped that the present collection will give many readers a chance to think about the role and importance of physics in the industrial world of the coming century.

Tokyo   E. Maruyama
May 1994   H. Watanabe
                                                                   Editors

Contents

PRESIDENTIAL ADDRESS

Yu. A. Ossipyan

Since three passed years the World where we work, engaged in our science, has greatly changed. Those were the remarkable years that have transformed the entire course of the World History by the end of the twentieth century. The disintegration of the Soviet Union has been the cardinal event, afterwards the bipolar structure of the World was gone.

This has led to the nuclear disarmament of the two Great Powers opposing each other - USA and USSR - coupled with their Blocs. And to major extent, it has reduced the terrible danger of the global nuclear war. Therefore, one of the prime troubles of the world-wide scientific community - the restriction - and then the discontinuance of the nuclear tests and the decrease the nuclear armaments level and then the total nuclear disarmament now ceases to be so acute and actual. I sense it rather directly as I personally together with the American colleagues have participated in the numerous negotiations and discussions on the subject. I would like to emphasize the fact that there are special permanently and jointly working Committees on International Security and Armaments Control (CISAC) in both the National Academy of USA and the Academy of USSR (now of Russia) representing these Great Powers in IUPAP. Their next meeting is fixed for late October in Washington.

Though the loss of the bi-polar structure of the nuclear competition by the World was of positive significance, but at the same time, it has brought the obvious difficulties and after-effects which became apparent in some areas in the form of aggravation of the regional conflicts which turned into the undisguised regional wars. (Yugoslavia, Lebanon, Tadjikistan, Georgia, Azerbaijan-Armenia and others).

The Scientific Communities and the intelligentsia from all the countries and nations should comprehend their most important role in terms of stabilization the situation within the many regions as well as the non-admission the development of the militant nationalism and fundamentalism. Unfortunately, the national elite among which in all the countries there are many representatives of the creative professions (science, art, culture) play a fatal role in the case of excitation of brute national instinct.

Hard economic problems emerging in many countries of the World have practically led everywhere to reducing the budget allocations for fundamental researches. This manifests usually sharp within the countries of the former USSR (Russia, Ukraine, Armenia, Moldavia and others), where the real political and economic crisis broke out, as well as in the countries of Latin America and Eastern Europe. This is accompanied by a loss of the public prestige of science and scientific education. This all takes place on the background of comparative rise in the cost in scientific research. We also need increasingly to employ the unique scientific instruments (accelerators, nuclear and fusion reactors, sources of various irradiation, cosmic media, telescopes, high-resolution microscopes (and so on). And to set up all these things we need to join the efforts of several countries.

However, the passed several years cannot be characterized only by these negative factors, during this period we have had essential achievements in physical science itself.

The results of the fundamental research, to a greater extent, have become the basis for the powerful developments of industry.   The results of the researches on solid state electronics, superconductivity, optical and electronic spectroscopy have led to creation of the new generations of computers, means of communications, storage, transmission and processing the information, devices and apparatus for medical technology, video and audio systems, and many others.

Studying into physics of strength and phase transformation and the other investigations involved in working out and in technology of the materials find an extensive application and are rather widely used in machine-building and aviation, in water and overland transport, and also in power-engineering and chemical technology. Chemistry and biology give a strong impulse to the technique used in daily living needs, as well as to agriculture and food industry.   The industry producing medicines and various biopreparations has extremely developed.   The scientific achievements of the recent time are responsible for these results.

Application of the scientific achievements for industry resulted both in a great progress in technique and in the substantive rise of the many non-polluting technologies should be especially pointed out.

All this has brought the enormous pecuniary incomes to the World industry, on the whole, and particularly, to big corporations.   On the background of these successes and enormous profits one can consider that the direct financial support for fundamental science is absolutely insufficient on the part of the corporations.

As it was earlier the basic financing for the fundamental investigations in the World is accomplished at the expense of the State subsidies.   In this connection our Commissions should brisk up its work for the purpose of attracting a bigger amount of money from the industry for direct finance of the fundamental science.   Certainly, it must be done simultaneously in search of the forms for organizing such support which excludes the direct pressure upon the fundamental science and the research projects on the part of the industrial corporations and their interests.   (Establishing International and National non-governmental foundations for fundamental science support).

Thus, side by side with our current activities in promoting science development and scientific education in accordance with our civic duty, we will have to participate in solving some critical problem within the contemporary human society.   In my opinion, for us, as the Scientific Association, the problems of paramount importance are the following :

1.OUTBREAK OF NATIONALISM AND REGIONAL WAR

If formerly, the problem N.1 was the fight against the global nuclear danger that consisted in coming out against the nuclear tests and in favor of successive nuclear disarmament, but now the problem is the struggle against the danger of the regional wars and the local inter nations conflicts.   Here our role is to create in each country and district the atmosphere of loyal democratism, tolerance and respect to somebody's opinion which is, in general, characteristic of the scientific approach.

2.   One ought to explain everywhere the necessity of big expenditures for science and education as these are the basis for prosperity both the whole countries and individual companies and individual people.   We need to have all this in mind without diminishing the efforts for progress and development physical science itself which is our paramount occupation.

I would be happy if in the course of general discussion here at General Assembly Meeting I could hear any comments on the considerations expressed by me.

Thank you.

General Assembly : W. Heinicke and Y. Yamaguchi.

Nara New Public Hall.

4

Academic Session : Yu. A. Ossipyan.

Tour to Horyuji Temple.

(Photo : Etsuko Kasagi)

INTRODUCTORY REMARKS

Hiroshi Takuma
Institute for Laser Science
University of Electro-Communications
1-5-1, Chofugaoka, Chofu-shi, Tokyo 182
JAPAN

First of all, in behalf of the Organizing Committee, I would like to extend the warmest words of welcome to all the participants, especially to the invited speakers who kindly agreed to present valuable materials based on the prominent academic activities of their own, by fully understanding the intention of the Organizing Committee. I am very glad to have such sessions where academic subjects are discussed in the program of the General Assembly, which is mostly spent for business. The reason is that the General Assembly is an exceptionally good opportunity to have very best invited speakers from a broad range of physics.

Then let me explain the reason for us to decide the title of the present academic sessions. I think one of the most prominent aspects in the physics of the Twentieth Century is its very close relationship with industrial technology. Physics gave a great impact to the advanced technology and the experimental technique in physics is ever progressing in the light of modern technology. In consideration of such a fact and also recent IUPAP policy of keeping close contact with industry, we have chosen this title : "Physics and Industry".

In inviting speakers of this session, we considered appropriate geographical distribution. An exception is that two speakers are invited from Japan from consideration that many of the over-seas participants might be interested how Japanese physicists participate in the industrial activities.

Let me spend a few minutes to cite one typical example on lasers, which is my specialty. Laser is cited frequently as one of the most significant inventions in the Twentieth Century. This is because the laser is a light source which emits light having a completely different nature. The laser beam is more like a very short wavelength microwaves rather than just a powerful light bulb. The reason comes from that the light emitting mechanism in lasers is different from classical light sources. Such a drastic concept came from a very strong desire to investigate the detailed structure of the microwave spectrum of ammonia molecule due to the nuclear spins and quadrupole moment. Such structures, generally called as the hyperfine structure, is so fine that they are normally hidden unresolved in the spectral linewidth. The absorption line observed in normal gas is very much broaden by Doppler broadening, and we must observe spectrum on atomic or molecular beam in order to reduce it. The problem in observing microwave absorption spectrum on molecular or atomic beam is that we cannot handle many molecules and naturally we loose in intensity. C.H. Townes overcame this difficulty by introducing an innovative idea of inverted population. By the use of a quadrupole electrode, he succeeded in separating the excited-state molecules from the ground state molecules, realizing a new state of matter, which is called inverted population or negative temperature.

This idea was extended to the optical frequency and created lasers. As you see in this example, an innovative idea frequently stimulated by a strong desire of a scientists who want to observe something new.   Such motivation seems to be stronger than desire to make money.   There are many other examples.   Reactors, the heart of the nuclear power plants, was developed by E. Fermi based on the advanced work at that time on slow neutron collision against uranium atoms.

Those are examples in which physics created new technology.   There are also a number of examples where the advanced technology stimulated the progress of physics. Does such a close relationship between physics and industry continue also in the Twenty-First Century?   Or does it have features different from what we experienced in the Twentieth Century?   This is a session to discuss such a subject.

In closing my introduction, I would like to express my thanks to all the speakers and participants, expecting active and valuable discussions.

FULLERENE AND FULLERITES - NEW MODERN MATERIALS

Yu. A. Ossipyan
Institute of Solid State Physics, Russian Academy of Sciences
142432 Chernogolovka, Moscow District, Russian Federation

Abstract. A review is presented describing quantum-chemical simulation and experimental data comprising the invention history of molecular clasters built of 60 and of more carbon atoms - fullerenes as well as a new form of solid carbon - fullerites.
Crystal structures and structure phase transitions taking place in pure fullerites and fullerites intercallated by alkaline metal ions and other ions, atoms and molecules and described. High temperature superconductivity of intercallated fullerites.
Application of the Bardeen-Cooper-Schrifer (BCS) theory to explain the superconductivity nature in fullerites. Phonon mechanism of electron pairing and Raman spectra of fullerites. Effect of high hydrostatic pressures. Mechanical properties are practical application of fullerites.

1. Introduction

The discovery of a new form of pure carbon - giant molecules called fullerenes and subsequently of a new crystalline form of carbon - fullerite crystals - has been a full - scale scientific boom over the past few years. Hundreds of laboratories all over the world are being engaged in synthesizing and studying fullerenes and fullerites and their derivatives, the number of publications amounts to two thousand, and the rate and scope of researches goes on growing.

This report is not a scientific review and it is not my aim to establish scientific priorities. This is rather a scientific popular lecture that better fits in with the spirit of this session. In view of this, not to overburden my report, I shall not make individual references in the text and figures since, to be exact and consistent, the number of such references must be very large. At the end of my lecture I shall give references to several recent very good reviews devoted to individual problems of fullerene physics and chemistry. The reader will find the necessary references to originals in these reviews.

2. History

The existence of giant molecules of carbon, boron or silicon has long been hypothized. Individual quantum chemical calculations evidenced for the possibility of stable A_{60}, A_{70}, and so on clusters. I know about the results of such calculations made and published in Moscow back in late 60s - early 70s. Possibly, there were other similar considerations and calculations, however, the ideas and models contained in them were hypothetical and based on general though quite sensible, from the stand point of quantum chemistry, assumptions.

It was only in 1985 that Richard Smalley's group at Rice University, Texas in collaboration with Harry Kroto at the University of Sussex, England, reported the result of their experimental studies that vapors formed upon laser vaporization of graphite

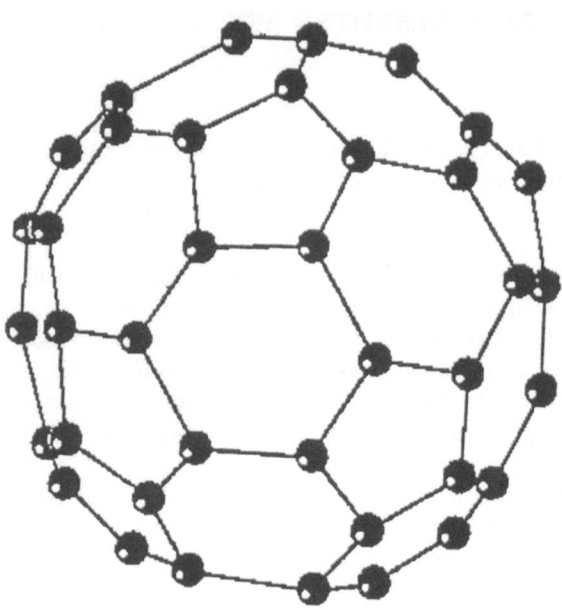

Fig.1. Fullerene C_{60} molecule.

contained considerable amounts of carbon clusters C_{60}, which was recognized mass-spectrometrically.

Concurrently, W.Kratschmer from the Max Planck Institute for Nuclear Physics, Heidelberg, Germany and Donald Huffman, University of Arizona, Tucson and their students L.Lamb and K.Fostiropoulos were concerned with producing "carbon smoke" by vaporizing a graphite electrode in an arc discharge in order to model the carbon state in interstellar dust. Their spectroscopic efforts have shown that "carbon smoke" contains C_{60} clusters which may well contribute to IR spectrum of interstellar dust. But they asked for more. Having collected the graphite residue resultant from spark erosion at the arc discharge, they dissolved it in benzene and, after evaporation of the solution, they got tiny crystals of C_{60} and C_{70} mixture, being first man made crystals of solid fullerenes-fullerites.

This one of the most cited works of late has opened avenues for laboratory synthesis of macroscopic amounts of fullerenes crystals, stimulated extensive activity in this field in many laboratories all over the globe. It has opened the new era of various physical and chemical experiments with these novel materials. Today a standard technology for synthesizing fullerite crystals involves the following stages : graphite vaporization in an inert gas atmosphere, one of the arc discharge electrodes being a graphite rod, collecting the vaporization products residue from chamber walls, dissolving this residue in an organic solvent, chromatographic column separation of solutions of various fractions

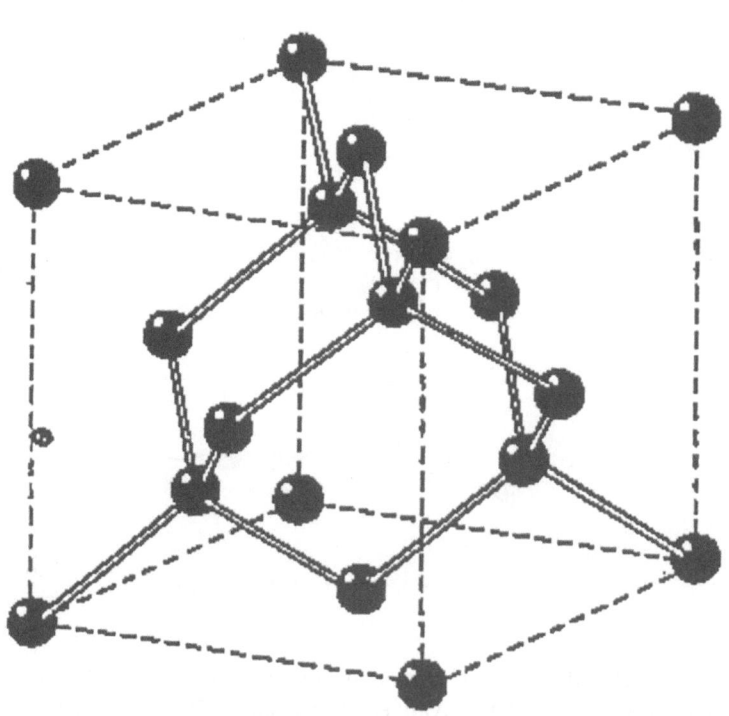

Fig.2. Diamond crystal structure.

from each other. The solutions are then dried to yield microcrystalline particles (powders) of various C_n fractions and, finally, these particles are sublimed to yield C_n crystals, it appeared that in the fullerene family C_n may be very large - more than 900, however, most stable and abundant are C_{60} and C_{70} the molecular structure of C_{60} is shown in fig. 1. This beautifully symmetric molecule belongs to the icosahedral symmetry is formed of 12 pentagons and 20 hexagons having the same edge dimension close to that of C-C bond in graphite. So, now we know three forms of solid carbon - diamond (fig. 2), graphite (fig. 3) and fullerite (fig. 4). A special interest to studies of physical properties of fullerites was aroused by publication of the sensational work of the group from AT&T Bell Lab's, who reported an observation of high T_c (=10K) superconductivity in fullerite specimens treated in potassium vapors. This phenomenon was attributed to the process of potassium atoms intercalation into the crystal lattice interstitial sites of fullerite, like it takes place with graphite. Further games with variation of the intercalant's type (of alkali metals) resulted in an increase of T_c up to 30K which, presently, is second only to T_c values of high temperature superconductors on the base of copper oxides.

It should be pointed out that the general idea of the molecular structure of fullerenes, similar to that of aromatic materials, has suggested that fullerenes and fullerites should be inert, chemically inactive materials. This, however, is not the case. It has been

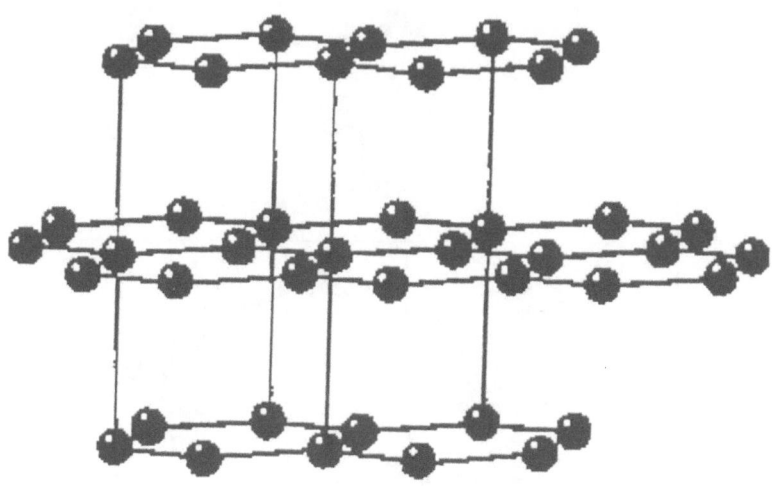

Fig.3. Graphite crystal structure.

demonstrated that fullerenes can participate in numerous chemical reactions with the formation of various chemical derivatives of fullerenes. This gave birth to a new broad field of organic chemistry which, undoubtedly, is very promising.

So, two approaches and, correspondingly, two different scientific fields can be distinguished in the fullerene and fullerite science :

first approach is to regard fullerites as stable formations that let atoms of small radius - mainly alkali atoms - intercalate into their interstitial positions. In this case the chemical bonds remain intact in fullerene molecules. An investigation of the physical properties of such alkali-doped compounds may be called the fullerene physics of today ;

second approach concentrates on possible chemical reactions involving break-down of bonds inside the fullerene molecules and formation of derivatives. This province is the fullerene chemistry.

Let us discuss these approaches at greater length and consider the principle results obtained.

3. Fullerite physics

An important division of this science is structural analysis of fullerite crystals. It has necessitated the development of special structural methods related to so-called Rietveld refinement. This method, used in powder diffractometry, is based on the acception,

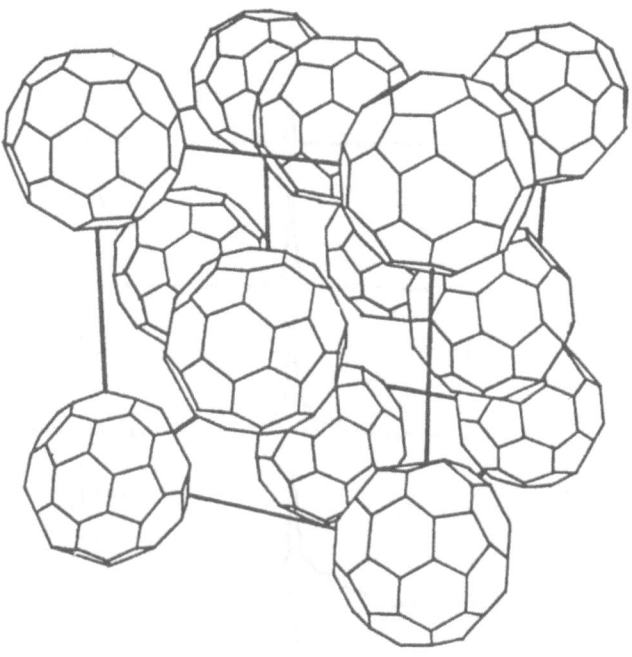

Fig.4. Fullerite C_{60} crystalstructure

from rational considerations, of apriori structural model and its subsequent correction so that the positions of all the diffraction lines were coincident with those derived from model calculations. To define reliably structures of various phases of pure and intercalated fullerenes this method is combined with a method of integral intensity measurements of different X-ray diffraction lines, with Laue methods to study single crystals, with neutron diffraction and electron diffraction methods.

1) Pure fullerite C_{60}

At room temperature a pure C_{60} has a crystalline FCC structure (fig. 4), but herewith the molecules are orientationally disordered due to their rapid quasi averaged spinning. In the X-ray time scale all the four molecules of the cubic cell are structurally equivalent. The lattice parameter $a = 14,16 Å$ comprises the value of the van der Waals molecular diameter $D = 10 Å$. At cooling below $T_m = 249\text{-}260$ K the phase transition to a simple cubic lattice (SC) is observed. This transition was documented by different structural methods - powder, on single crystals (X-ray), and, also, neutron and electron diffractions. Differential scanning calorimeter method confirms that the phase transition is a first order. Temperature dependence of C_{60} molecules spinning characteristics was studied in detail using the NMR on ^{13}C isotope. It was shown that above 140K there was one more phase transition, leading to narrowing of NMR lines (fig. 5).

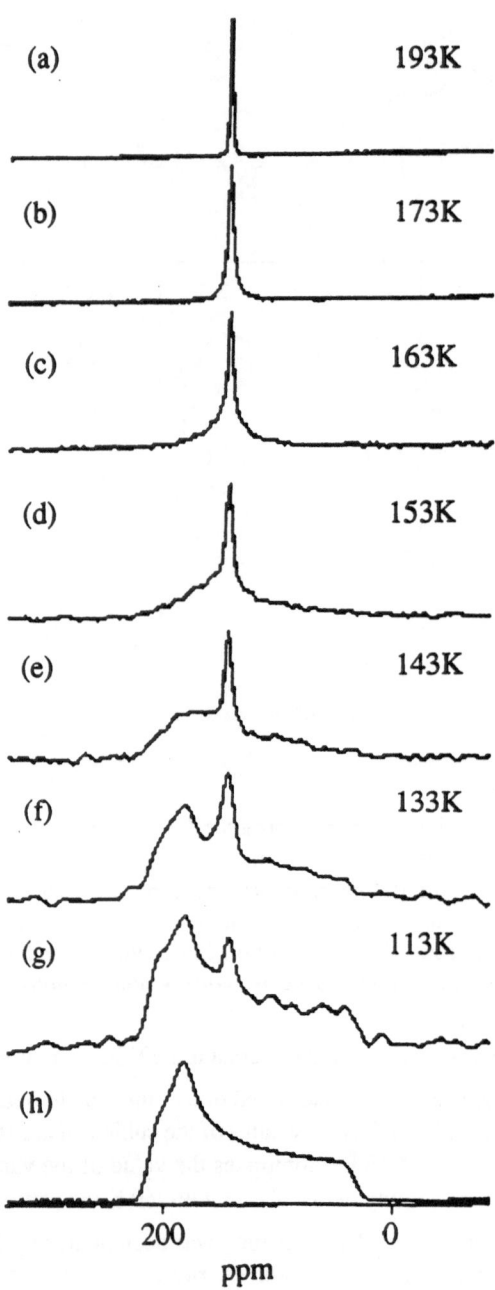

Fig.5. ^{13}C NMR spectra of solid C_{60} at indicated temperatures (a-g) h-tetramethylsilane.

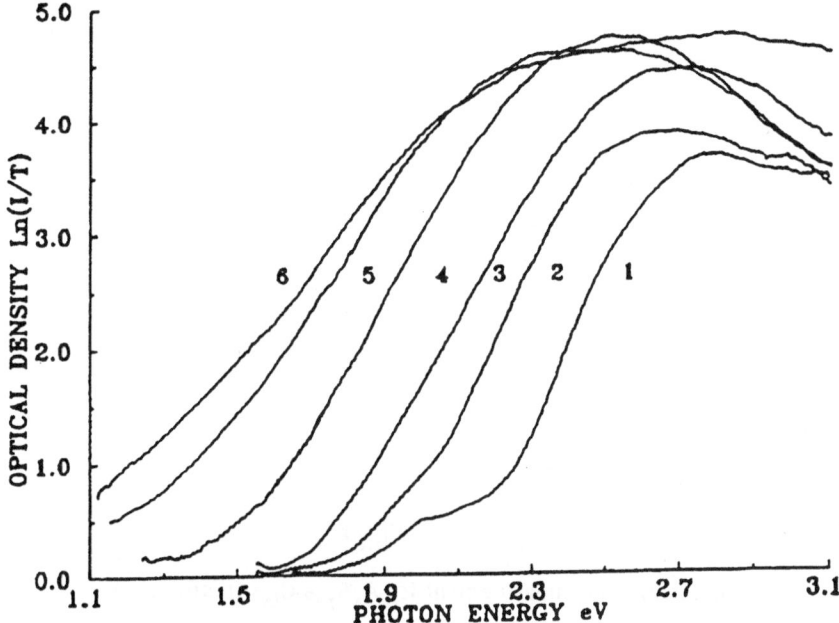

Fig.6. Absorption edge spectra of C60 crystal at T=300K and pressure
up to 20 GPa curves 1-6 correspond to pressure 0, 0.9, 3.1, 9.5,
14, 20. GPa respectively.

Effect of pressure. As the hydrostatic pressure is increased, the self-absorption edge
moves to the red spectrum side and the absorption edge shape changes (fig. 6).   The
compressibility modulus of C_{60} was defined from the displacement of the X-ray
diffraction lines position under pressure.
   2) Intercalated fullerenes
   As it has been mentioned, an interest to these materials was essentially stimulated by
observation of high-T_c superconductivity in C_{60} specimens annealed in alkali-metal
vapors (fig. 7).
Diffraction and morphology studies of C_{60} specimens aged at elevated temperatures in
alkali-metal vapors have shown that such treatment gives rise to different phases in the
system $A_x C_{60}$ where A (Na, K, Pb, Cs).
The following isolated phases have been documented :
$A_2 C_{60}$ - insulating
$A_3 C_{60}$ - conducting
$A_4 C_{60}$ - insulating
$A_6 C_{60}$ - insulating

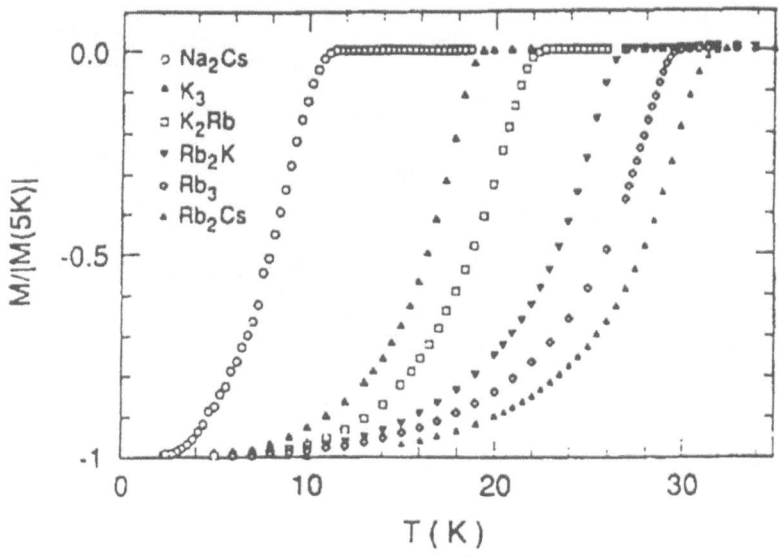

Fig.7. Shielding measurement for $A_x A_{x-3} C_{60}$ (normalized).

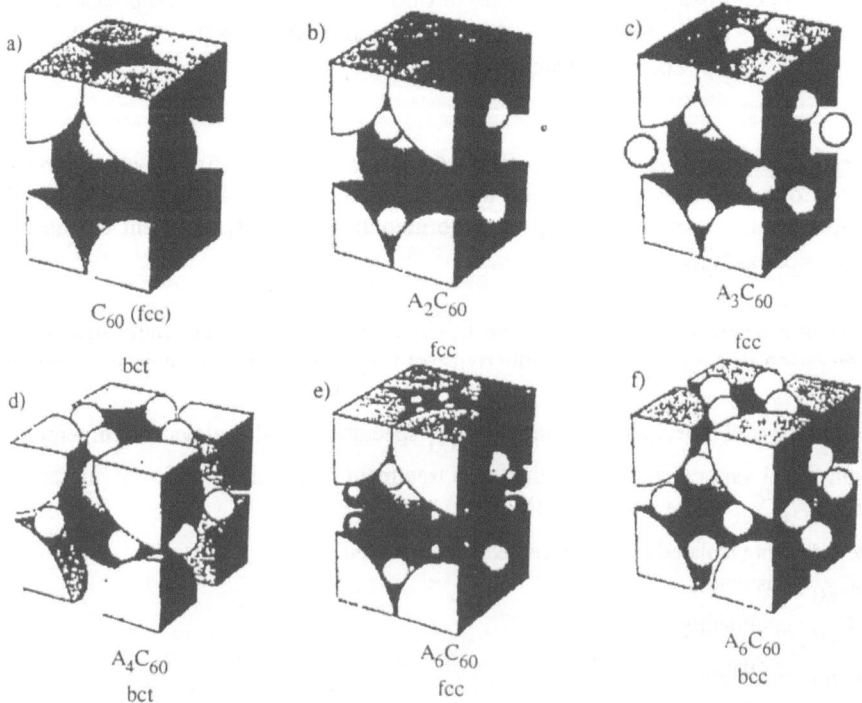

Fig.8. Schematic structures of C_{60} and $A_x C_{60}$. C_{60} - large spheres,
A - small spheres

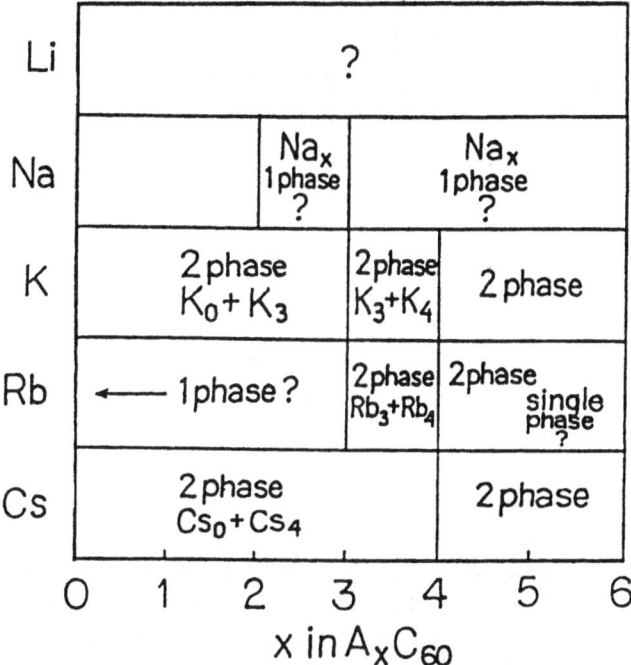

Fig.9. Proposed phase diagram for A_xC_{60}

Structural fragments of these phases are shown in fig. 8 where large balls stand for fullerene C_{60} molecule and small ones for alkali-metal atoms. Fig. 9 depicts schematically phase diagrams for C_{60} doped with various alkali metals. All the diagrams are seen to be of the same type, they exhibit clearly stoichiometric phases, and the intermediate compositions correspond to two-phase mixtures.

Of course this is a low-temperature part of the diagram of state. The questions of how these phases decompose or melt at elevated temperatures as well as the questions concerned with the value of their mutual solubility are to be quantitatively specified. It has to be noted, too, that ternary combinations of the type $A_xA'_{3-x}C60$ (A and A are different alkali metals) have already been synthesized. In order to specify the regions of existence of such phases one has to construct ternary diagrams of state. For some of them the phase structures have been already defined and it has been shown how A and A atoms get distributed in octahedral and tetrahedral interstitial positions. It is precisely for the double intercalant $Rb_2Cs\ C_{60}$ that a maximal $T_c = 31.3K$ is observed.

The calculation of the sizes of octahedral and tetrahedral interstitial sites and their comparison with alkali metals atomic radii-in order to construct models of their rational arrangement - are based on the structural models which, usually, the van der Waals radius is taken to be

$$R_w = 5.01\ \text{Å}$$

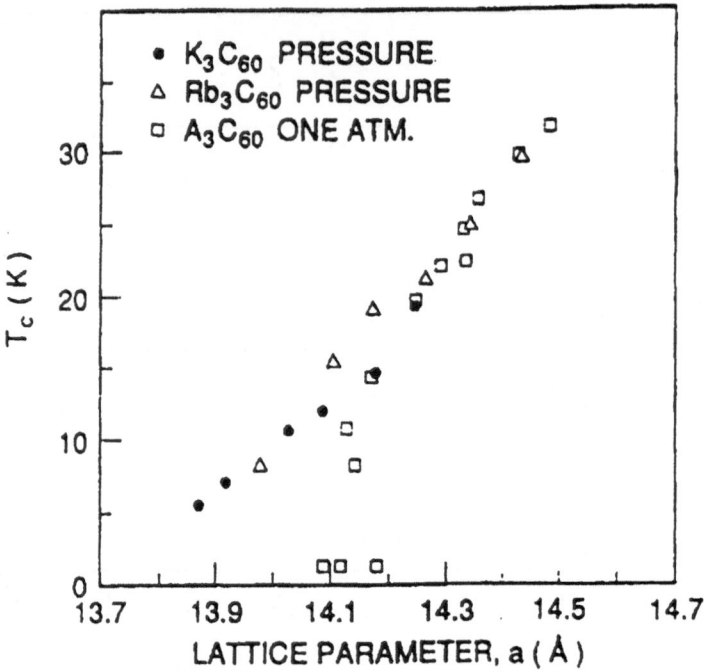

Fig.10. Relationship of the superconducting T_c to the unit cell size of A_3C_{60}.

■ A_3C_{60} - normal pressure

■ and △ - data from K_3C_{60} and Rb_3C_{60} under pressure

These calculations have already been performed both for real and hypothetic structures (BBC, BCT). Along with calculations of the exact position of diffraction lines (Rietveld refinement) this necessitated a great body of computation and creation of special computer programs (EASY/PULVERIX, NRCVAX) and others. The obtained results are rather promising, they suggest that the quantitative theory of formation of principal physical properties of intercalated fullerenes may be created earlier than it can be done for other multicomponent systems.

<u>Conductivity and superconductivity of alkali metals fullerides</u>   (A_xC_{60}) were studied most comprehensively for the system K_xC_{60} where, precisely, superconductivity had been discovered, and then the phase K_3C_{60} was identified as superconducting.

Later in a series A_xC_{60} other superconducting phases were found and, particularly, the triple ones of the type $A_xA'_{3-x}C_{60}$ having T_c 30K, which, presently, is exceeded only by compounds on the base of copper oxides. As it has been shown in fig. 7, Tc varies from 10K for Na_2CsC_{60} to 31K for Rb_2CsC_{60}. It turned out that the T_c value depends explicitly on the lattice parameter of intercalated fullerites, i.e. on the atomic volume of cations penetrated into the fullerite lattice (fig. 10). The T_c value is seen to increase as

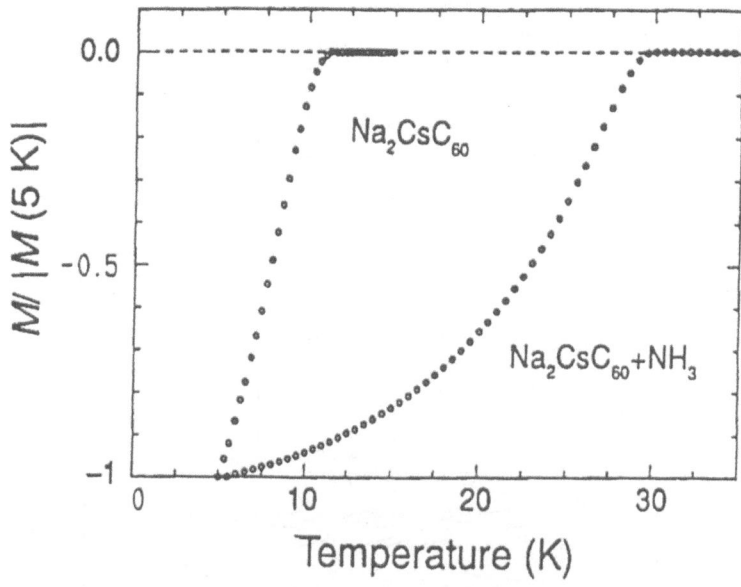

Fig.11. Normalized d.c. magnetic susceptibilities of Na_2CsC_{60} and $(NH_3)_4Na_2CsC_{60}$ measured in a field 2,5 Oe.

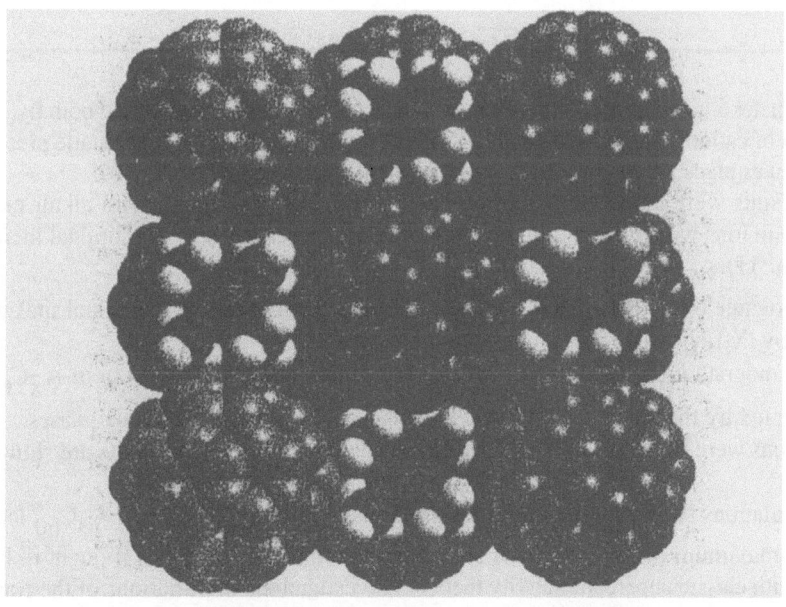

Fig.12. A model of $(NH_3)_4Na_2CsC_{60}$ with ordered $Na(NH_3)_4$ tetrahdra the C, Na, Cs, N and H atoms are represented by grey, red, green, blue and white spheres respectively

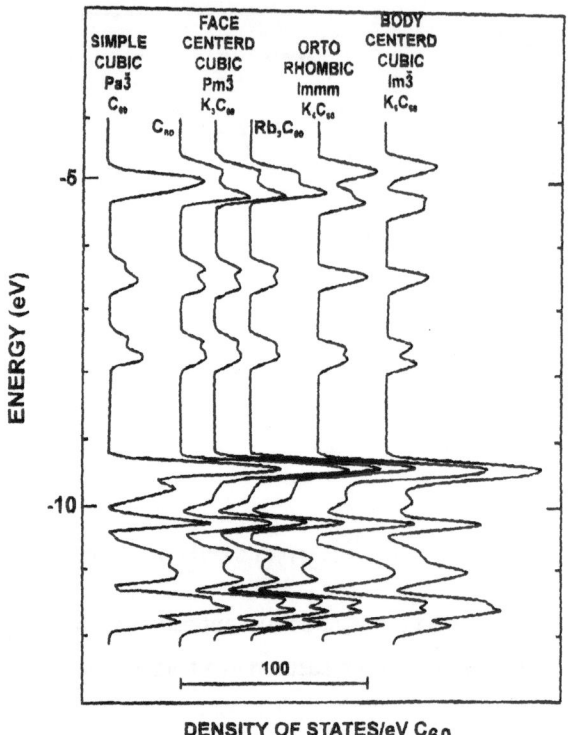

Fig.13. EHT band structures of A_xC_{60}

the parameter a is increased. Importantly, that a change in a can be attained both by variation of cations combination and by superposition of an external hydrostatic pressure. As this takes place, all the points fit well into the general regularity (fig. 10).
These results were the base for important experiments when along with alkali metals ammonium ions were used as intercalating ions. which readily led to additional increase of T_c (fig. 11).

The structure of the obtained phase and the arrangement of ions in interstitial sites were defined by X-raying (fig. 12).
Further understanding of the character of particular phases in the system A_xC_{60} is corroborated by the calculations of the band electronic structure of these phases. The calculations were based on the Extended H kel Theory (EAT). The results are shown in fig. 13.
The calculations show that the Fermi level of the superconducting phase A_3C_{60} locates near the maximum of the density of states of the conduction band. as it ought to be in accord with classic superconductivity theory. Investigations of variations of the Raman spectrum of C_{60}, as an alkali metal was being intercalated, proved very useful.

Fig. 14 illustrates schematically individual Raman spectrum regions with indications of what vibrational types are responsible for a particular spectrum region.
Observations of the K_3C_{60} superconducting phase formation have shown that the high-

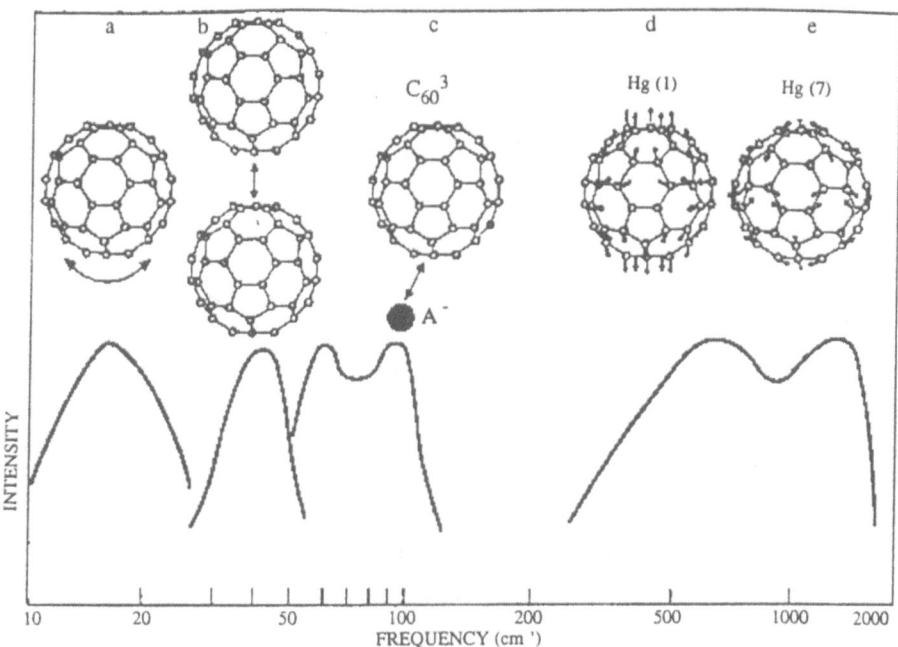

Fig.14. Various vibrations in the A_3C_{60} compounds can contribute to electron-phonon coupling and may be important for super-conductivity.

frequency spectrum region changes during this event. Then as the potassium content increases and the K_6C_{60} phase forms the spectrum changes again (fig. 15). These data suggest the conclusion that the occurrence of superconductivity in the K_xC_{60} system is related to the electron-phonon interaction in vibrations of the Ag(2).

Signigicant role of the electron-phonon interactions in the process of current-carrier coupling in the K_3C_{60} system is also supported by the presence of isotopic effect for T_c observable in K_3C_{60} at substitution of ^{12}K by ^{13}K isotope (fig. 16).

These data bear witness for the fact that electron-phonon interactions play a significant role in the mechanism of the occurrence of high-temperature superconductivity of intercalated fullerenes. They also raise hope for consistent construction of quantitative physical theory of this phenomenon [1,3].

4. Fullerene chemistry

Investigation of chemical reactions which involve fullerenes and of the properties of these reactions products is a vast and rapidly developing field of today's organic chemistry. Since this audience consists mainly of physicists, it would serve no purpose to go into

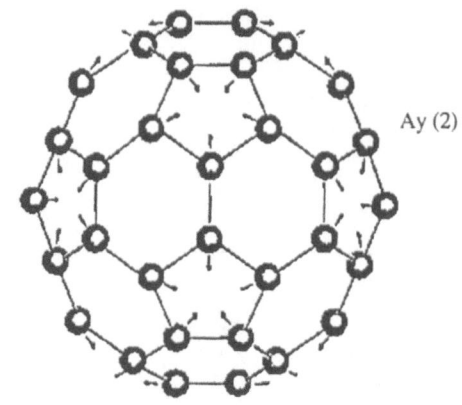

Ay (2)

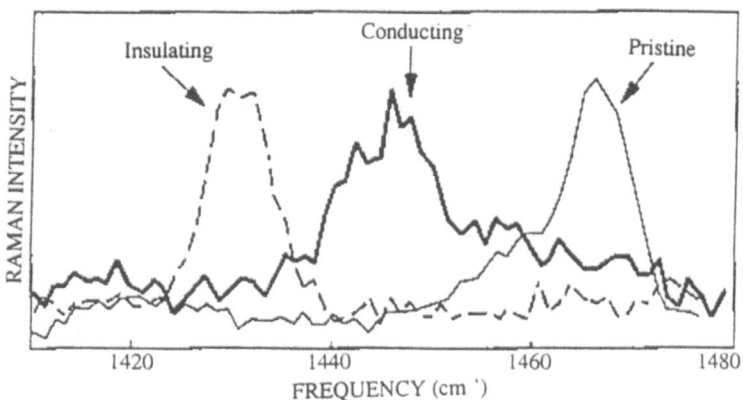

Fig.15. Raman spectra evience for charge transfer from potassium
atoms to the C_{60} molecules.

the heart of the chemical problems, I shall, therefore, only briefly list them. The people
interested in these questions I can refer to the excellent review by Roger Taylor and David
R. Walton published in Nature in the summer of 1993 [2].
Roughly speaking, all fullerene compounds can be classed into three categories :
1. Intercalated compounds wherein fullerene molecules in the crystal lattice sites retain
integrity and identity whereas foreign atoms occupy interstitial positions in the lattice.
2. Endohedral clusters obtainable upon capturing of a non-carbon atom inside a fullerene
molecule (encapsulation). In this case the fullerene molecule also retains its structure.
3. Exohedral solids formed from fullerenes to which foreign atoms or molecules are
covalently bonded on the outside of the carbon cage.
We have already discussed the first type compounds. I can only add that intercalating
atoms may be not only alkali-metal atoms (cations), but, also, e.g. iodine atoms which,
in these processes, probably manifest themselves as anions.   The $C_{60}J_4$ phase was

found in the $C_{60}J_x$ system, its structure was examined by x-raying and characterized.

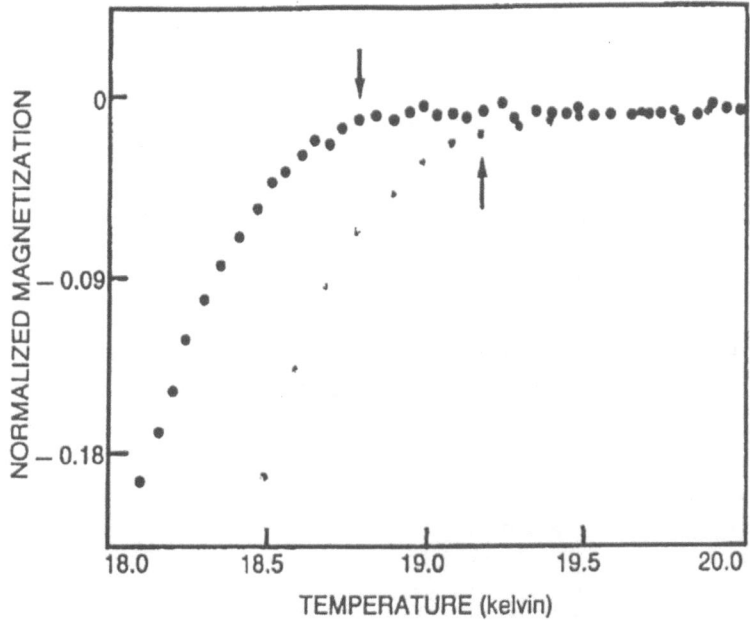

Fig.16. Isotopical effect. Magnetization transition in isotopically pure
$K_3{}^{13}C_{60}$ occurs at the temperature 0,4K lower than in
$K_3{}^{12}C_{60}$

Another interesting C_{60} - derivative intercalated compound is tetrakis - dimethylaminoethylene (TDAE) of the formula C_2N_4 $(CH_3)_8$. Despite a large number of different atoms in the molecule all of them have a comparatively small atomic radius and molecules as a whole can locate in the interstitial sites of the C_{60} structure. This compound, is specifically, a ferromagnetic, having the Curie temperature of about 16K, which so for is the highest among organic ferromagnetic. Seemingly, a partial charge transfer from intercalant's molecules (or atoms) to fullerene's molecules is of importance for structure stabilization in all intercalated compounds.

As for endohedral cluster solids, then we know, so far, the La (a) C_{60} compound, where the symbol (a) implies that La atoms are inside the cage. There were other attempts to synthesize endohedral cluster solids with other rare-earth atoms inside the cage, however, low quality of specimens and inadequacy of experimental material do not make it possible to draw final conclusion about the crystalline structure of these compounds.

An investigation of the chemical reactions leading to the formation of the third type compounds i.e. exohedral ones as well as of the structure of these compounds is, precisely, the province of the fullerene chemistry comprising the main ideas and approaches of organic chemistry. The key moment in the understanding of the fullerene molecule behavior in various chemical reactions is that the occurrence of the double bond

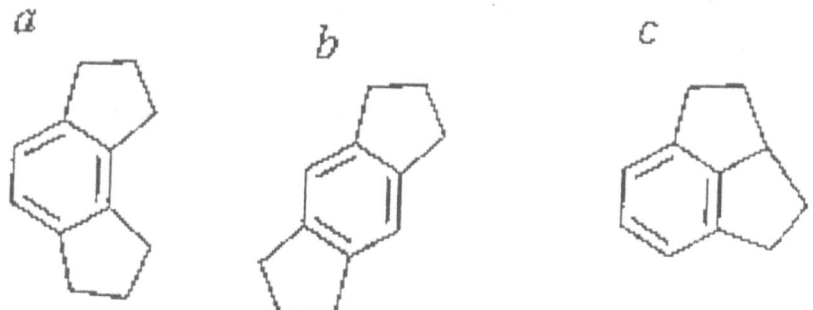

Fig.17. Possible disposition of two pentagonal rings adjacent to *a* hexagonal ring. Disposition *b* and *c* introduce instability.

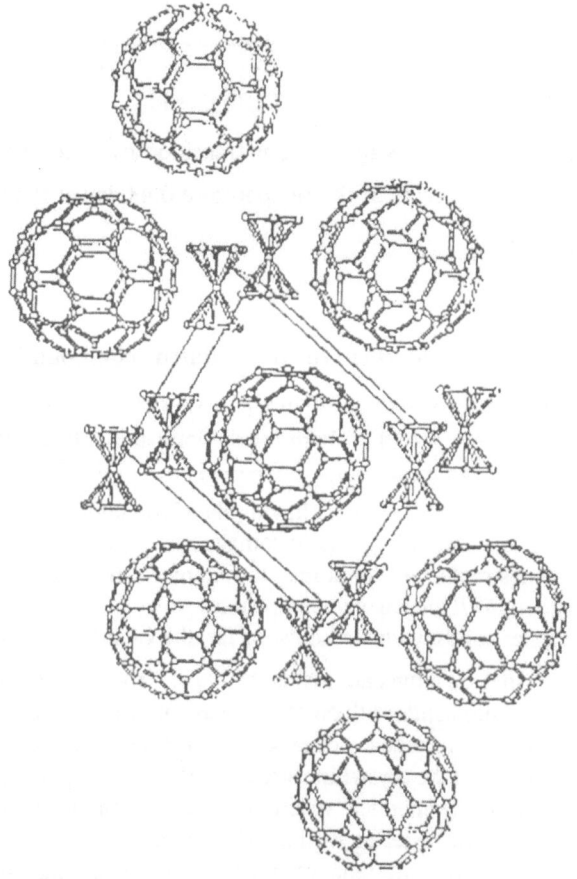

Fig. 18. Host-guest structure of C_{60} (ferrocene)$_2$.

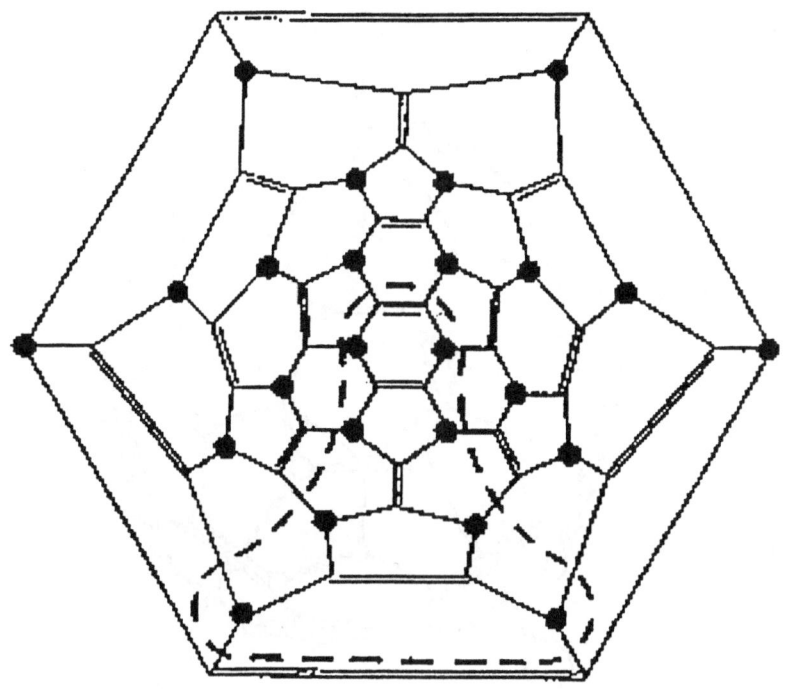

Fig.19. Schlegel diagram showing the 24 non adjacent sites in C_{60}

in pentagonal ring must be excluded.   There is only way of packing pentagons and hexagons so that a stable isomer could be formed.

Some versions are illustrated in fig. 17.   As contrasted from aromatic molecules, fullerenes do not possess atoms of hydrogen or other added groups, - therefore, they are not capable of substitution reaction.   Substitution reactions can take place only with derivatives, especially those formed addition.   The electronic structure of fullerenes molecules suggests that they ought to have an increased electron attracting.   This governs their chemical behavior, for example, they readily react with nucleophiles.

At a slow crystallization from benzene C_{60} fullerene molecules yield solvates, $(C_6H_6)_4C_{60}$ in which spinning of the molecules is so slow that it is possible, using x-ray diffraction method, to define the structure of the single crystals.

The same results are obtained at crystallization from cyclohexane.   There are some other complexes from which co-crystallization with benzene occurs.   All these materials obtained at co-crystallization exhibit so-called host-quest structures, an example of which with ferrocene is shown in fig. 18.   Here is much in common with intercalated compounds when structural stabilization occurs due weak interaction related to charge transfer.

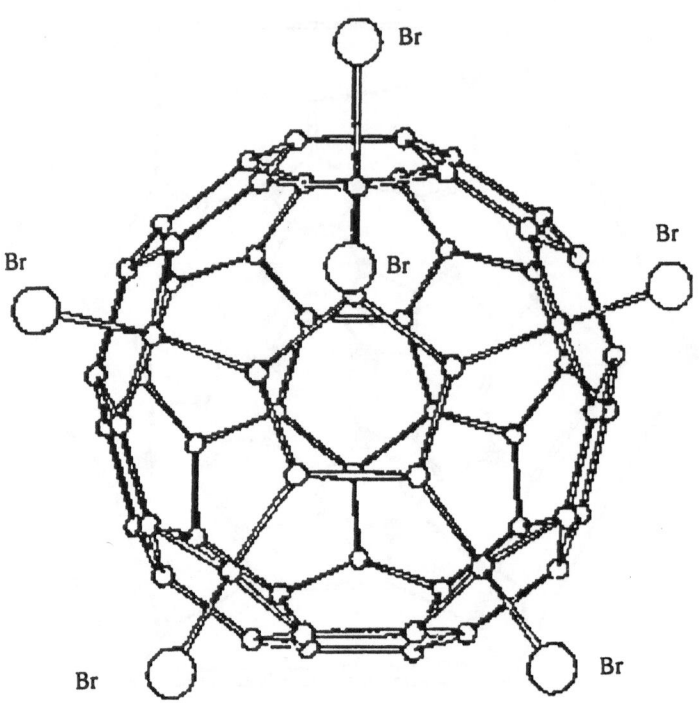

Fig.20. Structure of $C_{60}Br_6$

Analogous structurs are obtained at the interaction with sulphur ($C_{60}S_{16}$ and $C_{70}S_{48}$). These are formed of S_8 rings.

As for remaining typical chemical reactions of fullerenes, known by the present time, I shall restrict myself to their brief listing.

Anion formation and oxidizing processes.
These processes are of clear electrochemical nature. In particular, fullerene catalyzes oxidation of H_2S to S. This process is rather typical for organic chemistry and it is being intensively studied using platinum asacatalyst. The interaction of C_{60} with t-butyl-lithium belongs to the same class of reactions. Details of such reactions are rather complicated.

Addition reactions
These reactions can be categorized into three groups :
1) Cycloadditions
2) Additions involving bridging
3) Additions of separate groups

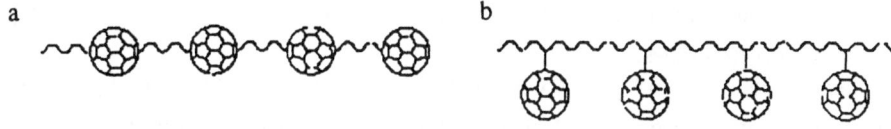

a

b

Fig.21. 《Pearl necklace》 and 《pendant chain》 polymers that in principle can be made of fullerenes.

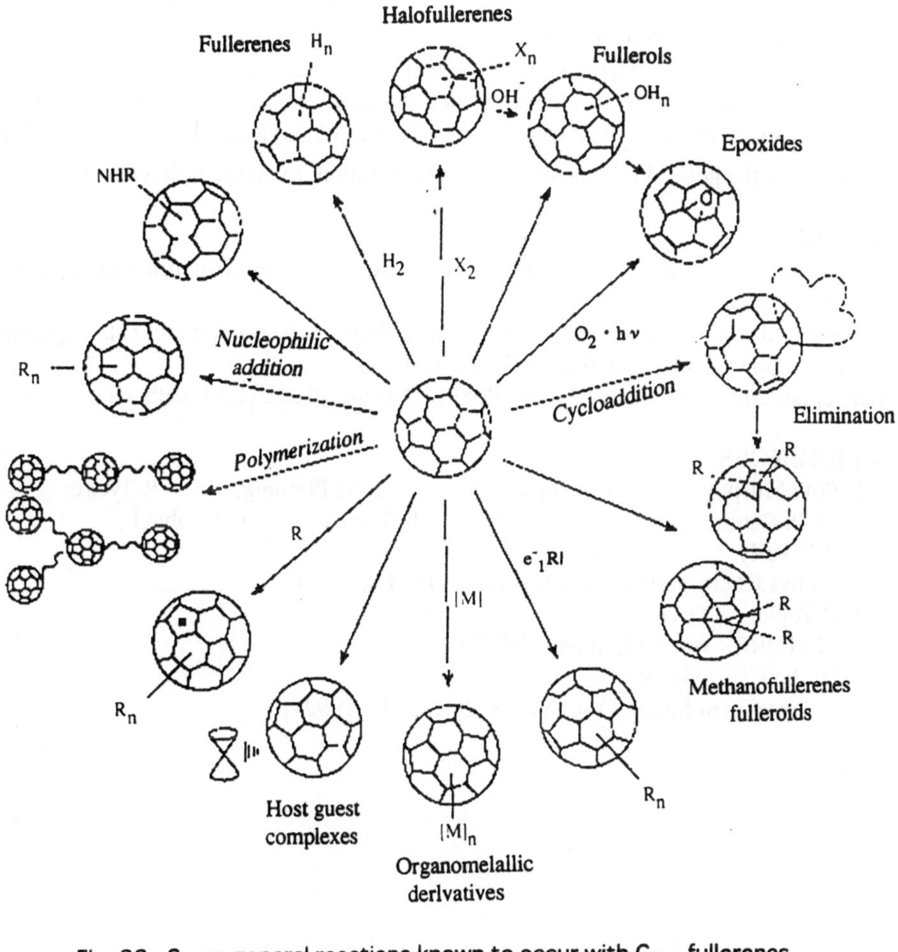

Halofullerenes

Fullerenes H_n

Fullerols

Epoxides

NHR

H_2 X_2

R_n

Nucleophilic addition

$O_2 \cdot h\nu$

Cycloaddition

Elimination

R

Polymerization

R

$e^-_1 RI$

|M|

R

R

Methanofullerenes fulleroids

R_n

R_n

Host guest complexes

$|M|_n$

Organomelallic derivatives

Fig. 22. Some general reactions known to occur with C_{60} fullerenes.

Category 2 comprises reactions with the formation of epoxides from isolated C_{60}. These are oxygen bridges.

Addition of methylene to C_{60} and C_{70} goes via the formation of carbon bridged. There are recent reports on synthesis of methanofullerenes. In reactions with metals than as aromatics. Among reactions of addition of individual groups one may distinguish addition of halogens and hydrogen. Only 24 groups can be added to C_{60} so that two of them were not neighbours (fig. 19).

Specific structure of $C_{60}Br_6$ is shown in fig. 20.

Polymerization

Polymers comprising C_{60} may be confined to three types : two of them are the pearl necklace type, fig. 21, third type is pendant chain, fig. 21. One may assume that their two - and three - dimensional versions may be described as a polymer net work or lattice. Probably, the third type having a direct bond between the cages from at polymerization of C_{60} under the ultra-violet irradiation in the absence of oxygen. In conclusion of this section I give the general scheme of all chemical reaction involving fullerenes (fig. 22).

Conclusion

It is clear that besides a great scientific interest investigation of fullerenes promise considerable practical implementation.

In addition to what has been already mentioned, they can be used in solid state quantum electronics, optics and production of electric batteries.

Many opportunities for chemical technology and chemical analytical methods are in sight.

REFERENCES

1. D.W.Murphy,   H. J.Rosseinsky,   R.M.Fleming,   R.Tycko,
   A.O. Ramirez,   R.C.Haddon,   T.Siegrist,   G.Dabbagh,
   J.C.Tully,   R.E.Walstedt,
   J.Phys.Chem.Solids., vol.53, N11, 1321 (1992)
2. R.Taylor,   D.Walton,
   NATURE, vol. 363, p.685 (1993)
3. O.Zhou,   D.E.Cox,
   J.Phys.Chem.Solids, vol. 53, N11, 1373 (1992)

PHYSICS AND ITS ROLE IN TECHNOLOGY*

P. Chaudhari
IBM Research Division, T.J. Watson Research Center
P.O. Box 218, Yorktown Heights, NY, USA

Abstract.   This talk is divided into three parts. In the first part we show the connection of physics to a variety of industries. In the second part we discuss the changing enviroment of research and development world-wide and, in particular, in the US. In the third part we describe the management of science in an industrial laboratory, such as IBM.

It's my plan to talk about two of these three items. I shall talk about physics in industry and then a little bit about the future research and in particular about a term that I believe will become increasingly important. It is called search. I will not have time to talk about management of science, although I will touch upon it.

What I intend to do is to take a number of examples of the role of physics in a diverse set of industries. All of these examples, I'm sure, are known to you but it is sometimes useful to look at them at one sitting so you can appreciate the effect that your discipline has on technology.

What I have done here is to list human needs on one side and then taken a particular branch of physics that I'm active in, which is called condensed matter physics, and show you the impact that this particular field has had on human needs. I will not go into historical examples. In fact, I will only use examples that are currently of active interest in the sense that these examples, if successful, will lead to the next generation of Technology. But I will acknowledge the past as I have done here.

If you start from the first one, which is food, you'll find that solid-state physics, at least, has had no direct impact. Perhaps in packaging of food it may have played a role, but I do not consider that a direct impact. On the other hand, in medicine, if you go outside biological physics, you'll find that in almost all diagnostic tools that you can see today solid-state physics, or physics in general, plays a very important role. You are, of course, very familiar with X-Rays from the days of Roentgen, but perhaps not so familiar with MRI's development and that of SQUID and so I will show you examples of that and then I will return to this chart.

The idea of MRI, of course, stems from nuclear magnetic resonance but another area of physics it draws upon is super-conductivity. In MRI one is looking at slight changes in relaxation rate of the proton from that in pure water to the water that is in contact with tissue, diseased on healthy. Those slight changes in relaxation rates show up as signal which you display. I'm sure all of this is very well known to you and you can get beautiful images of human anatomy, as shown here. Very clear distinction between tissue and bone is obtained. What is changing or perhaps what is evolving, is not so much a refinement of this technique

---

*   Written version is excerpted from the oral presentation, which
    contained many more illustrations.

but an attempt to combine these static techniques with dynamic ones. In other words, what you would like to do is to be able to say, "This part is diseased but the area around it is not diseased" and you'd like to say that by examining the patient's response while you look at the image and the surgeon can say "this part is still healthy, so as I operate I should not be touching it."

Here's an example. It combines MRI with a technology that's been around for some time called SQUID Technology. It relies on the magnetic field that's generated as a electrical signal goes down our neurons. As you very well know, if you have a current going down a wire there's a magnetic field associated with it. In a similar fashion, if you have a neuron there and there's a pulse propagating down this, you get a very, very small magnetic field. What you would like to do is to sense that. To do this you rely on a phenomenon called super conducting quantum interference. I will not try to explain it other than in a hand-waving way. Basically it's a loop which has what are known as two Josephson elements across it, and if there's a magnetic field threading the loop you get a phase shift around it and, depending on the magnitude of the phase shift, an appropriate voltage signal. I've shown you these oscillations here (Fig. 1) and their modulus is the flux quantum. The flux quantum is 2×10^{-7} Gauss per centimeter square so that through a centimeter sized loop you can sense extremely small magnetic fields. It is this property that one relies on in trying to use SQUIDS as a diagnostic tool.

People have built such diagnostic tools. A semi-commercial device is shown here. This is available and has been used and it's been used in combination with MRI as I show you here. This is a tumor in the head which is clearly identified by MRI, that's this particular dark spot. What the surgeon was interested in was how big a hole should he make as he operates on it. So they combined this with SQUID technology and as the person moved their hands and their feet this particular area of the brain would light up which is then mapped with the magnetic field that emanates and therefore when combined with MRI gives you these maps, shown here in red and green and, now, the surgeon knew what exactly to avoid. I show this only to indicate to you the trend that the physics that we are involved with is leading up to. This is, of course, SQUID technologies done at Helium temperatures and right now people have demonstrated that the high temperature SQUIDS can also be used to look at brain waves. So you have here a natural development in physics; a discovery leading to applications.

Let me come back to this first view-graph which I'll go back and forth with. I think it's certainly well known that, if you're thinking about shelter, which is another very important human need, construction materials are very important. This has, of course, been konwn to us since time immemorial. In fact, entire ages of civilization are known by these materials. More recently the emphasis has been on composites. These composites, while new to us as mankind, are certainly not new to nature and you and I are excellent examples of composites. Composites in nature come at all length scales and we talk about length scales in our science but nature uses them all the time. I show you a particular example here. The reason, of course, nature uses it is because it needs different properties combined together to perform a set of different

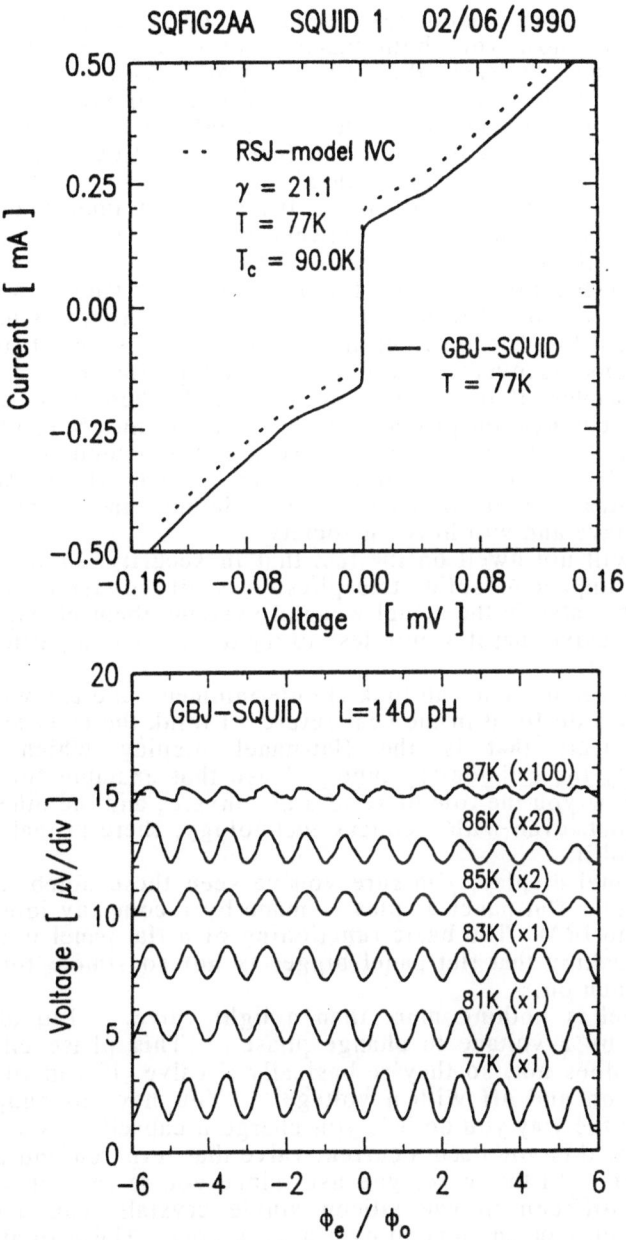

Figure 1

functions. We use composites for exactly the same reason and I have two examples to show you; both taken from biology mostly because they provide to me, at least, the spirit of man. You may notice this man running but he has no natural legs. The set of requirements from our legs that we take for granted have to be emulated. For example they have to take the impact, the torsion, your feet have to flex, have to twist. All of these have to be integrated inside a material or a set of materials to give you the desired functionality. It's also very important to understand what happens at that interface between the body, or the body fluids, and the metal or the material it interacts with.

This has been a big issue and I show you this particular view graph for two purposes. One, I want to focus your attention on this ballerina. This ballerina had lost her hip joint that was replaced by an artificial hip joint. The surgeons discovered that with time the hip joint loses contact with the hip. However, if they corrugated it and the human tissue, as it grew around the corrugation provided a very solid joint. She, of course, now uses that to dance freely. I also use it for a second reason to tell you that if you're interested in materials there is a movie available from the National Academy of Sciences which depicts the wide impact that materials have and will have on society.

I think I will not dwell on the fact that in security, in weapon systems, there, the impact of solid-state physics is all-pervasive, not just in the materials but also in the actual weapon systems themselves. I just think it's so pervasive that it's pointless to try to give you a particular example, so I'll skip it.

On the other hand, one can look at entertainment and ask what will solid-state physics do for it in the near future? I think the most simple example to demonstrate that is the flat-panel display which is replacing increasingly the cathode-ray tube. I use that example for two reasons. One, to show you the role of solid-state physics and the other to illustrate the point that, even with mundane technology, there is real physics to be done behind it.

The flat panel display, I'm sure you've seen these notebook computers. this particular flat panel display is made by a company jointly owned by Toshiba and IBM. The basic functioning of a flat panel is rather simple. although making that flat panel bigger is not so simple for reasons that I'll just touch on.

A flat panel is nothing more than a light valve. Liquid crystals are controlled by a voltage to change phase. This phase either transmits light or it does not, so they're basically a valve, if you like. You turn that valve on and off with a voltage. You have to supply a voltage source and the way you do it is you charge a capacitor as you put current in it and for this you need a current valve that turns on and off and that's a transistor. In this case, you use--since you're reading large displays and it is difficult to use silicon single crystals--you use amorphous silicon which you can deposit over a large area. These light rods as they open and close go through color filters, the three primary colors, and that combination gives you the display that you see.

Now it seems like a fairly simple technology and you might ask yourself "well, why don't we make these into very large area displays today?" The reason is actually very simple. The mobility of amorphous silicon is currently limited and we don't understand why. So you see, here is this technology that's limited by this very simple physical question. Why is it that amorphous semiconductors have an intrinsic low mobility? If you could improve the mobility by a factor of ten, that would be a significant advance in this particular technology because it leads to a very large display. You realize that I'm simplifying the issues here but that's basically whats holding this technology from going to large areas.

This interplay between technology and science is always present. Solid-state physics, of course, has always played a very important role in communication and computers. In fact, another way of saying it, without that particular field there would be no computers and no communication in the modern sense of the word.

What I'm going to do is to take you very quickly through a set of these examples, again, things that you must surely know. These kinds of pictures you've seen many times and, as you can see, in today's technology, half-micron lithographic prepared features are fairly common. People have made transistors in the laboratory where the line width is 100 angstroms or thereabouts and these transistors also operate. So we know there are no fundamental laws which say that you cannot continue to miniaturize silicon technology for the next ten, fifteen years.

However, we are also discovering that there are natural length scales. We all know them in physics, they lead to the notion of dimensionality. We're finding that a large number of these devices now operate in the regime of one, or two, or zero dimension. We're trying to understand what does this mean is terms of device technology, or does it lend itself to new device ideas? But the important point that I want to leave with you here is that because of this high level of integration, the cost per bit of information has dropped dramatically. This is a point I want you to remember because I'm going to come back to it later when I talk about the implications of this in future R&D activity.

Although you have heard a great deal about silicon technology, perhaps you don't know as much about the disc that is present in all your computers, and, which you take for granted. On this spinning disc has on its surface a detector that's flying typically at about 1,000 angstroms above it. Mind you, it's doing this while you're shaking or jerking your machine! In order to detect a higher density of information, this slider has to come even closer to the disc and you have to use more sensitive techniques.

You find that, right now, there's a great deal of emphasis about using giant magneto resist sensors. These are these multilayer materials or two-phase materials which have very large magneto resist signals. They were discovered only a few years ago but are already being used in technology. The point is that the time span between discovery and technology is shrinking and is not, as some people believe, staying constant or is about 15 years as the canonical number. Sometimes, however, it can take longer for several developments have to occur together.

For example, there is another technology, which has a different history to it, and this is the optical read-write disc technology. In this particular instance you can read or write the information and what you're relying on is the fact that you can induce magnetization to point up or down by using a laser to warm up a local area in a biased magnetic field and then you

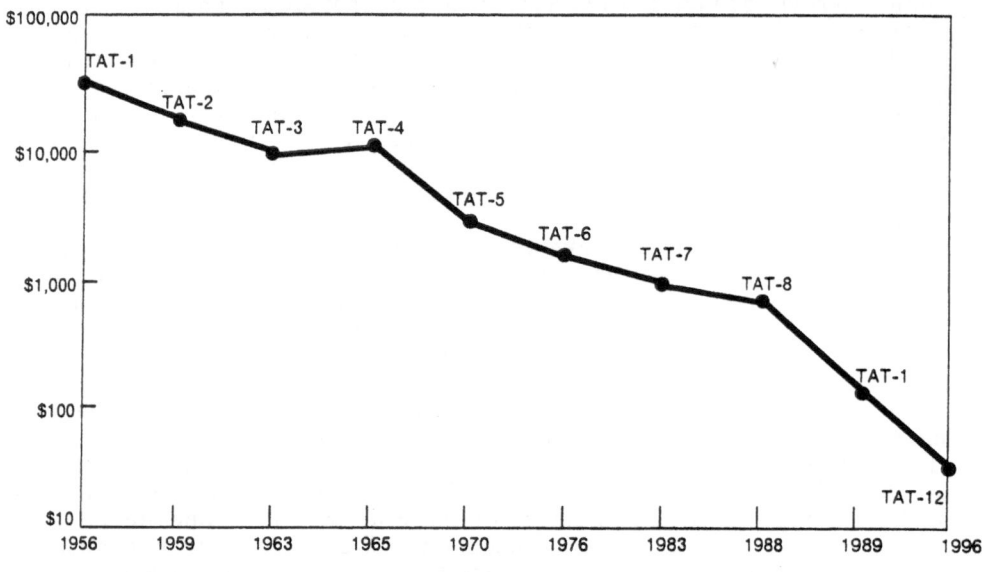

Figure 2

rely on Kerr rotation to sense whether the polarization is up or down.
This technology has been evolving over twenty years. It was first
invented in the mid-70's... the materials were invented in the mid-70's
then it required the advent of personal computers and the availability of
lasers, that you can buy now so inexpensively. This technology's needs
has forced us to try to look for shorter wavelength lasers. So here's an
example where technology is forcing us to invent solid-state blue lasers.
The shorter wavelength, leads to a factor of four, roughly, in the packing
density of information.
This is a view graph from AT&T about their transatlantic cables, and I'm
sure I can get similar view graphs from other telephone companies, which
tells you the fiber-optic cables they're laying worldwide. These fiber-
optic cables with their wide band are, as you can well imagine, going to
change the way we communicate with each other.
That is a second point that I would like to focus your attention on. The
first was computation and now communication. These will change the
way we do R&D. As in semiconductor memory technology, there is a
similar curve: the cost-per-circuit-per-year for, say, trans-Atlantic
communication, has decreased by four orders of a magnitude over the last
four decades, almost paralleling what you see in computers (see Fig.2
from AT&T). So both the cost of computers and the cost of
communication have come down dramatically over the last four decades.
This, more than anything else, is going to change the way we operate and
do research in the world. That's the point I'll come back to.

Let me try to end this first part quickly now. I will not talk about energy because I'm sure Morrel Cohen may touch on this in his talk. Let me end with just showing you a couple of view graphs in transportation. A car, as you very well know, is nothing more than a set of materials, of a very diverse kind, all put together to carry you safely in an economic, efficient and comfortable way. We're used to this idea and we take it for granted. A car is an excellent example of man-made composite, if you like, on a macroscopic scale. This trend in trying to optimize cars is going to continue but, of course, what is going to change is that the amount of electronics in these cars is going to go dramatically up. In fact, it will be the biggest single change and these cars will communicate with each other, will communicate through satellite with central stations. Essentially, a car will be much like your home, with all the gadgets in it, so you can find out where you are, find out where your wife is...perhaps you don't want to find out where your wife is, or your wife to find out where you are...but in either case, cars will change dramatically in the future.

I want to show you this one because my trip is still very fresh in my memory. You have probably heard about this plane that's going to go from New York to Tokyo in a matter of a few hours. I'm looking forward to that because I just took this journey that took almost 23 hours from the time I woke up to the time I arrived at Osaka Airport. For this plane to fly, you'd have to develop a whole set of solid-state physics dealing with the structural strength of materials at very high temperatures; not only for the efficiency of the engines that propel it but also the fact that the skin of the aircraft itself has to take high temperatures and, therefore, has to have structural integrity if you have to fly it safely back and forth.

I have many other examples but I will not show them now. I want to simply remind you that there is a very solid connection between physics and industry. So you have to ask yourself, if this connection exists, and it is certainly bound to exist in the future, what is going to change? Well, I've already told you what is changing is the fact that the communication between us, between laboratories, between research groups, is now almost instantaneous. We know what's going on in another laboratory if we care to find out. You also find that the number of researchers worldwide...I realize in some countries it's coming down, in others it's going up more rapidly...but if you look at the aggregate number of researchers worldwide you'll find it's increasing. So the research worldwide activity is increasing.

This environment makes a very dramatic difference in how you organize research in the corporate lab. Let me try to illustrate that. I think it's a point that particularly people in the United States and Europe have to come to grips with because it's forcing them to change and it's not a change that is due to bad management practices of the corporations. It's a change that is very much an evolutionary change. They are going much more from, what I would call, research towards search, whereas other countries are going much more from search towards research. I'll try to make that clear.

What I want to do is first show you that the knowledge of what we consider critical technologies is now pervasive. All of us know what the critical technologies for the next generation are.   I'll show you a list and you will recognize that all of you know what it is.   The details of what I need to develop within that critical technology is also well-known and it's important to understand that.   What this means is the different laboratories are working all, roughly, on the same set of ideas.   Since different laboratories will be emphasizing different things, through chance and through emphasis and through a variety of reasons, there will be new developments in different laboratories around the world.   No one laboratory can dominate anymore.   This means that I, as a researcher, have to recognize that not only must my research lab develop knowledge that it needs but that I must be ready to search for this knowledge where ever it's available.   That's what I mean by search.

So that the idea for corporate lab such as an IBM, which is very big, to do everything we need to do is an obsolete idea.   What it really means is that, as a corporate lab, I have to downsize to a size that I do enough research so as to recognize what I need and from where to get it, I will tap them and then I will try to bring them rapidly into my own corporation and use them for my products.

In the United States and in Europe that's a relatively new thought.   In Japan that was not a new thought when you started, say 20 or 30 years ago, you went out and brought the ideas in.   So you're increasing in research and less in search.   The rest of the world, particularly the United States, is realizing that research is now not as needed as much as it was needed, say, 20 years ago because it's now going on all over the world.

That's a painful period of adjustment, for it feels like the money for funding is going down.   It's not because of bad management, I think it's just an evolutionary process we're seeing.   Let me show you again by examples what I mean by that and then I'll talk about this temporal contraction.

Here is a list of technologies that were prepared by a panel set up by the U.S. Government and mandated by Congress (Fig. 3).   It's supposed to come up every two years with a set of critical technologies and provide that technology list to Congress and it's supposed to be done by The White House.   This is the list we came up with early this year and we listed there our nine, what we call "technologies that are considered essential for future development."   We then asked ourselves a question, if you think these are important what do some of our trading partners think about them?

We found that you, in Japan, find the same nine technologies to be important and you're putting emphasis on them.   We then asked ourselves, what does the European Community do?   And once again you find, barring one or two exceptions, they too are putting additional money in the same set of things that the United States Critical Technologies Panel considered essential.   Even within the European community there are countries that are putting more effort in it than the average of the European Community.

We thought this was interesting but how active is this and how do you measure it?   So we decided to look at the patent activity because that's usually a fairly important thing to do if you're in industry and wish to protect yourself.   And here is the patent activity in the same nine areas by

Technology Intensive Industrial Sectors

	USA	Japan	European Community	France	Germany
Materials Synthesis and Processing	●	●	●	●	●
Electronics: Micro- and Opto-	●	●	●	●	
Software	●	●	●		●
Distributed Computing and Networking	●	●			●
(Flexible) Integrated Manufacturing	●	●	●	●	
Applied Molecular Biology	●	●	●	●	●
Transportation	●	●	●	●	●
Improved Electricity Supply	●	●	●		●
Pollution Minimization and Remediation Techniques	●	●		●	

● = targeting activity announced by named country

From the Report of the Second National Critical Technologies Panel (U.S.)

Figure 3

different countries, and what I want to draw your attention to is not just the set of expected countries, U.S., Japan, Germany, United Kingdom, France, but you begin to see Taiwan and Korea becoming active in the same nine areas (Fig. 4).

Clearly, most countries in the world know that these areas are important for the future and, within that, they are trying to position themselves to aquire by research and search the requisite knowledge they need.

Here is another measure, just to beat the point home, and this is the direct investment by foreign countries in the United States in these nine areas. That is, countries outside the U.S. may recognize that the U.S. may have a particular strength in an area, so they go into the United States, open up a research lab and try to tap that skill in house. The United States, by the way, does the same so it's not just a one-sided thing.

What this implies, as I said earlier, is that this idea that a corporate lab can somehow keep its knowledge within its boundaries and give it out when it thinks it's important...this idea of captive knowledge is not valid any more, and certainly will not be valid in the future. As a result of that, management realizes that it no longer need the same size laboratory that it did say, 20 years ago. And so the laboratories tend to shrink in countries that are used to doing research over the last 40 years. It also means that research becomes a smaller fraction; much of the time is spent it trying to bring knowledge in, so the time required to develop the technology becomes a dominant factor and so you have, basically, what I call a temporal contraction in the sense of the total time it takes to bring a product out.

So what did I try to tell you? Let me repeat. All of it is, in my opinion, fairly obvious...and I apologize for giving an obvious talk...but it sometimes helps just to face the obvious. It is clear that physics will continue to play a role in industrial development. I've given you a few examples but I could have picked many other examples. So I think we should take that for granted...we should take it for granted that the people who pay our salaries, or pay our salaries indirectly, will continue to do so in the future and that the number of people like us in the world is going to grow.

We should also realize that those countries that are used to doing research will continue to do research, but they'll recognize that there are many, many players around the world. And, using modern communication and computers, they'll try to access it and position it so that they get the knowledge first. You'll do it, we'll do it, they'll do it, everybody's going to be doing it. Therefore, in any given country, the size of the amount of R&D activity will be optimized depending upon what the nation can afford and how they position themselves in areas they wish to play in.

Now if you're going to change from research to search in the United States, then the management incentives that you use have to be very different. In the past you might have argued that since he discovered it, he's the inventor" and so give him the reward. But now it has to be different. It just can't be that, you also have to give credit to the person that brings that knowledge from another lab. And this is in recognition that this person, from a corporate point of view, has played just an important role as the guy who invened because he allowed a product to go

Relative Patent Activity by Nation and Technology

	USA	Japan	Germany	United Kingdom	France	Taiwan	Korea
Materials Synthesis and Processing	●	◒	●	◒	◒	◓	○
Electronics: Micro- and Opto-	◒	●	○	○	◒	●	●
Software	◒	●	○	○	○	◓	●
Distributed Computing and Networking	◒	●	○	●	◒	◓	●
(Flexible) Integrated Manufacturing	◒	○	●	◒	◒	◓	○
Applied Molecular Biology	●	○	●	●	●	○	○
Transportation	○	◒	●	◒	●	○	○
Improved Electricity Supply	◒	◒	◒	◒	●	◓	●
Pollution Minimization & Remediation Techniques	◒	○	●	◒	◒	◓	○

From the Report of the Second National Critical Technologies panel (U.S.)

KEY   Number of Patents

Average Frequency	> 1000	100 to 1000	< 100
> 20% above	●	●	●
~ 20%	◒	◒	◒
< 20% below	○	○	○

Figure 4

out which then made money for the company. And that requires thinking from a management point of view very differently. How you organize yourself and what incentive do you provide.

In Japan it may be the other way around. You may, in the corporate labs, have played great emphasis, 20 years ago, on going and searching and bringing that knowledge back. But now you have to begin, as you do more research, to require a different set of management skills than the ones that required searching for that knowledge.

So I think an interesting evolution around the world and I think it's all caused by the fact that computers and communication are dramatically different from 20 years ago. It's not a question of being a small difference, it's a radical difference when I can pick up the phone, fax my message or send an E-mail message with both visual or audio information. So that is the way I think the world of R&D is going to evolve and each of us has to position ourselves in this new world in the right way so as to optimize our role.

PHYSICS AND INDUSTRY

Yves Farge
Pechiney
10, place des Vosges-Cedex 68 - 92048 Paris La Défense

Abstract.   The interaction between physics and industry, which plays a major role including in our daily life, is examined and analysed during the last decades. What it has benn. What it is. What it should be.

Introduction

To prepare this paper, a literature survey has revealed little has been written on the subject of the interaction between physics and industry over the past 15 years. It is astonishing because everybody believes that relations between physics and industry are particularly important and fruitful. Everybody knows that the first discoveries made during the nineteenth century changed our society ; for example thermodynamics has helped to understand the steam and combustion engines and the laws of electricity and magnetism gave birth to the electric motor ; human and animal mechanical labour has been replaced by these devices and have liberated a huge amount of human time. The electric light bulb has given light during the night and especially in winter, giving more time to humans to think or to work. Electricity and electromagnetic laws have given the telephone systems and all the telecommunication world, reducing the size of the planet and opening each individual to the entire world. At the same time, progress in metallurgy, internal combustion engines and gas turbines have increased by several orders of magnitude, the distance that a human could travel during a given time, reducing the size of the planet and greatly increasing the interactions between populations. Nuclear physicists have invented the thermonuclear bomb as well as nuclear energy production ; for the first time in history politicians had to work together from all countries to avoid a world disaster. More recently semiconducting properties of doped silicon (and some other semiconductors) have introduced a new social revolution with microprocessors and computers, replacing the boring part of the brain activity and once again liberating a huge amount of time for more constructive activity.
　　Physicists must be rather proud of this history. They know that a lot of industrial activities were developed from their discoveries with the creation of very large companies opening new jobs. "Physics had a remarkable record of contributions to business and industry"[1] . However, today, physicists are experiencing that the situation has changed and that their discipline is no longer the leading one and they are challenged by chemists and more and more by biologists who are in fact largely indebted to physicists for most instrumentation facilities. Scientists in all disciplines see that large companies, especially in the United States, are reducing drastically their fundamental research with a clear lack of confidence in its output. It can be observed since the sixties a slow-down of basic discoveries able to change our culture (microprocessor and laser essentially ; the high Tc superconductors as well as fusion are very far from industrial reality). In the same period a gap can be observed between civilian and military technologies, the latter

becoming more and more sophisticated, therefore more and more expensive, reaching costs which are too high for civilian applications. During this period the physics community has been more and more linked with military industries which have now well known problems. If chemists are well linked with chemical industrialists all around the world, the link of physicists with civilian industries is much weaker and seems to have declined since the sixties.

However, since the beginning of the eighties it appears that interaction between industry and physics is increasing again. Not for the development of totally new products and processes but largely to have a better understanding of existing processes, in order to improve them. Physicist are realising that industrial activity is raising very exciting questions in terms of fundamental knowledge and that relations between physics and industry is not a one way process (physics to industry) but must be a two way process which is not new since it is well known that thermodynamics was invented to understand the properties of the steam engine and not the reverse. This new trend is superimposed on a more permanent trend which is the use by industry of new experimental technologies developed by physicists.

In this paper I will try to describe the interaction between physics and industry through the main analytical methods, then some missed opportunities ; afterwards I will give my views on what I see as a come-back of physics to industry with some recommendations to improve this new interaction.

I. MAIN ANALYTICAL METHODS

Instrument developments have always been a large part of the research work in Physics. The last Nobel prize was awarded to Georges Charpak for its instrumental work on ionic chambers for particles detection. New useful instrumental developments are usually quickly adopted by industry which always need new methods to solve problems. Another example of a recent technology is the Atomic Force Microscope, new experimental method quickly adopted by industrial laboratories. Surfaces are of great importance in all industries (materials are always fixed by their surface etc.......). Surface analytical methods are widely used and AFM is a new method permitting observations impossible before. It has been used by Pechiney for metal corrosion studies to measure the depth and the shape of small corrosion points on aluminium alloys.

Many other instrumental methods are available to give structural as well as chemical information on surfaces (grazing incidence X-ray scattering, electron back scattering, photoelectron spectroscopy, Auger spectroscopy etc....).

Instrumental methods to study bulk properties of solid or liquids are very numerous : for example, X-ray and neutron scattering, electron microscopy. For polymers, NMR, ESR, Raman fluorescence spectroscopy as well as linear and circular dichroïsm are widely used. X-ray scanners as well as neutron radiography or ultrasound are extremely useful to identify defects in bulk materials. For example several X-ray pictures are taken of each turbine blade of aircraft engines to verify that they have no inclusions or voids which could reduce their fracture toughness.

Figure 1 shows vertical stresses in a rail slice measured by neutron scattering. A very well collimated white neutron beam defining the scanning zone is scattered and measured at a Laüe angle. A precise measurement of the wavelength and the angle gives a precise measurement

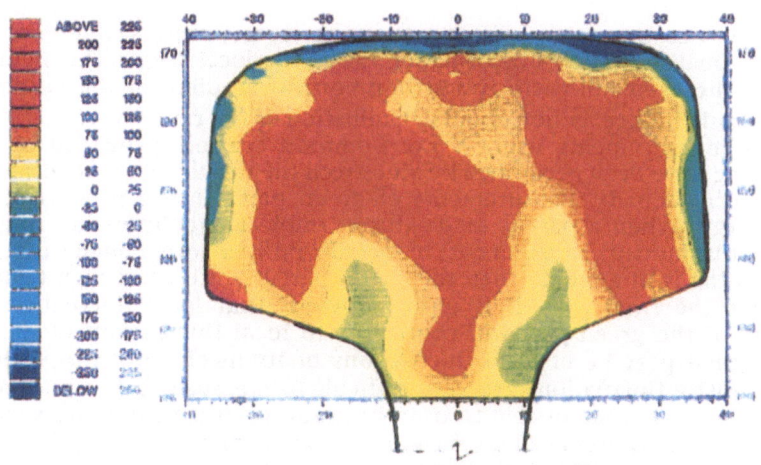

Z Stress in Unused 370 Rail Slice (MPa)

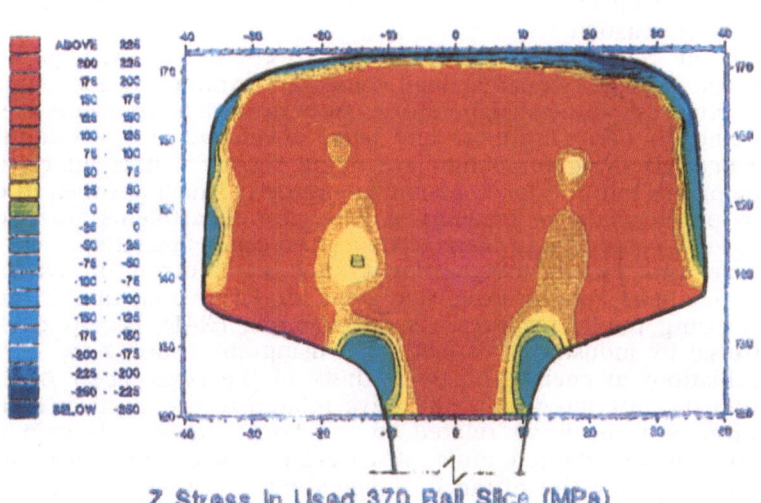

Z Stress in Used 370 Rail Slice (MPa)

by courtesy of Dr. Peter J. WEBSTER (University of Salford - UK) and ILL Grenoble

Figure 1

Residual stress contour (in MPa) of a rail slice through the steel lattice parameter, variations measured locally by thermal neutron Internal stresses are proportional to the lattice parameter variations. A reduction of the parameter corresponds to a compression and an increase to a tension (from PJ. Webster [2]).

of the lattice parameter whose variations from normal values are directly proportional to the local deformation, i.e to the local stress. This figure shows that in the used rail there is a strong vertical gradient of the vertical stress which could induce an horizontal delamination of the rail.[2]

Similar measurements using X ray Compton scattering instead of neutron scattering have been performed by Sintertech Cie (a subsidiary of Pechiney) to perform density measurements of green parts in the process of powder metallurgy. When a part is made (for example a synchronizing ring for the automotive industry) a green part of metallic powder with a binder is made by pressing it in a mold ; then this green part is sintered at high temperature to make the final part. It is quite obvious that local fluctuations of the density of the green part will contribute to local fluctuations of the density of the final part i.e of local fluctuations of its mechanical properties. Such local density fluctuations are very difficult to measure because a green part is very soft (is is impossible to use ultrasonic measurments) and very brittle. Compton scattering is ideal for such local measurements :

- Compton scattering cross-section is proportionnal to the number of electrons i.e to the local density,
- Compton scattering is a non destructive method,
- it is possible to explore given volume by a right collimation of the incoming beam and an adequate collimation of the scattered beam.

This method has been used to measure the density of each tooth of a synchronizing ring of a gear box before sintering and to understand the origin of the fluctuation.

Most of the related equipments are cheap enough to be acquired by a medium size or large company. Small companies cannot buy such equipment and they often ask academic laboratories to perform measurements they need ; this method is becoming more and more developed and will continue to develop because more companies are doing research. It is interesting for these companies but also for academic laboratories which are then in contact with new problems. Some academic engineering departments are offering a complete set of equipment to characterise a given class of material.

Very large and expensive equipment such as neutron or synchrotron radiation facilities are too expensive even for large companies ; therefore they are using public facilities. As an example, EXAFS is now a classical method used by industry : this method is using absorption X-ray edges and their oscillations at energies in the vicinity of the edge. These oscillations come from the interferences of the wave related to the electron excited by the photon and the wave related to the backscattered electron by the neighbours of the excited atom. A carefull Fourier transform of these oscillations is giving the distance of neighbours as well as their chemical identity (by the intensity of the reflection of the electron wave), the chemical identity of the excited core atom being given by the threshold wavelength. Synchrotron radiation with its continous spectrum in the X-ray range is particularly well suited for such experiments. This method can be applied in many systems, for example aluminium alloys doped with copper where it is possible to measure the environment of copper atoms. EXAFS has in particular greatly contributed to the field of catalysis. Figure 2 shows the L_3 absorption edge of platinum in alumina coated with Pt ; the red line is measured just after coating and shows isolated platinum atoms with oxygen neighbours[3]. After reduction by hydrogen at 350°c and 20 bars the spectrum is totally changed and shows only metallic platinum aggregates which are the active catalytic species. By EXAFS, it was then possible to optimise the best conditions of temperature and pressure for this reduction.

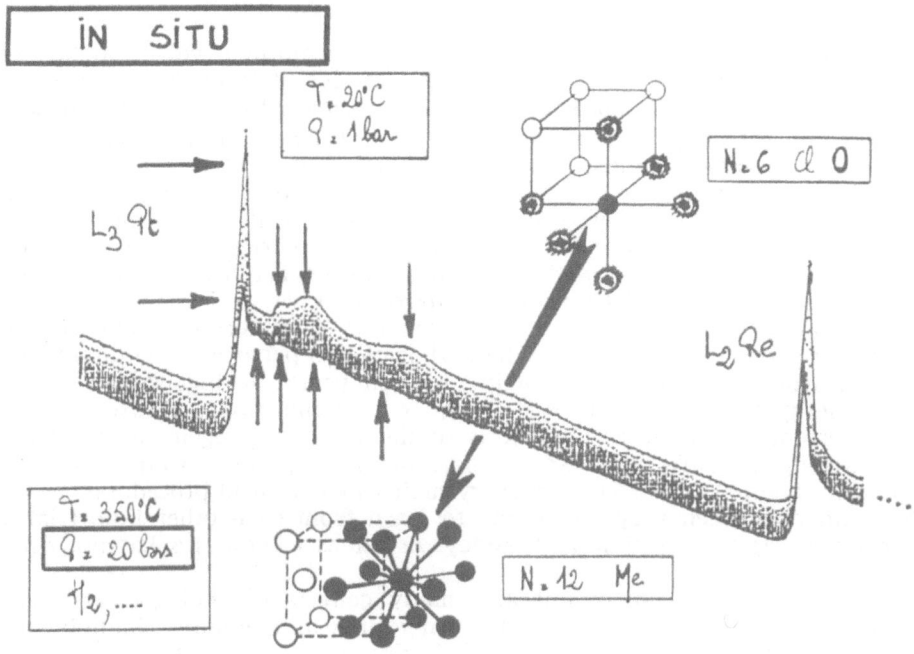

Fig. 2

L3 absorption edge in the X ray range adsorbed of platinum on alumina (EXAFS spectrum).
- red curve : platinium is oxydized and an analysis of the oscillation above threshold is giving up the distance Pt-O,
- blue curve : after reduction at 350° C under hydrogen atmosphere at 20 bars, metallic and catalytic platinum agregates are formed. The metallic Pt has an absorption very different from the oxyde.

It is also very interesting to notice that many companies making equipment using all these physical methods have been created by physicists as a spin-off of academic laboratories. Americans have been particularly effective in creating such companies, an example being Hewlett-Packard from Stanford. Today, the instrumental work in physics is not recognised enough and creation of new companies making instruments is probably slowing down. For good interaction between physics and industry, it would be worthwhile to recommend a better recognition for the physicist's career of invention in new experimental methods.

II. Missed Opportunities

Although new experimental methods are developing very quickly in industrial laboratories or plants for light physics, it is no longer true for heavy physics such as high energy physics, nuclear physics or astrophysics. In an interesting paper, Amendolia[4] has published an enquiry on the interaction between industry and physicists in the field of high energy detectors ; his conclusion is that the general trend is to make too much equipment in the high energy physics laboratories. This equipment could be more efficiently subcontracted to specialised industries which could then increase their know-how and apply it to other activities, while in the laboratories, more time could be devoted to research.

The alliance of fundamental research and engineering is one of the characteristics of heavy physics. The author of this paper was the director of French synchrotron radiation facility for ten years, having worked in the high energy physics laboratory at Orsay. It is astonishing to realise how the management of a large accelerator is similar to the management of a large industrial plant. To have an accelerator or a plant working with an efficiency higher than 90% requires very similar management procedures.

It would have been very interesting to learn from each other industrialist and physicists to save time and money. As far as I know this has not been done.

Not only the management but also many generic technologies are very similar. The process control of a modern rolling mill requires simultaneous measurement of many parameters with feedback on all available parameters, which requires a mathematical model of the mill, a mathematical treatment of the data and several computers to drive the mill. It is similar for any kind of accelerator. Here again more interactions could have been very useful, specially because high energy physicists were usually ahead in such given technologies which had to be improved for each generation of accelerators while similar technologies had to be improved for very large industrial plants. That has not been exploited ; while it would have been possible to train very good engineers who could later have been hired by industry.

Such training could also take place in many specialised technologies such as high speed acquisition and treatment systems, local treatment of data, ultrahigh vacuum, X-ray detectors, magnetism, adaptive systems etc..... Heavy physics to develop new equipment (for example modern telescopes with adaptative optics) must develop new methods which are very often ahead of industrial technologies. We can regret that the physicists who lead such developments are educated only for inhouse sake and that these major projects have not been enough used to train young students for industry.

Certainly, industry on its side was not really prepared to hire such engineers. Today, it is time to reduce the gap, to consider from industry that such big laboratories are excellent to train engineers to new technologies and, for these laboratories to consider that they are in the academic world and have training duties.

III. The Come Back of Physics in Industry

Interaction between physics and industry cannot be considered only from the instrumental and engineering viewpoint. Physics has a lot to say to Industry and a lot to learn from it. Today, each company has to face very intense competition on the world market. The profit of a company is usually few percent of its sales with an added value (production cost) of the order of half of the sales and prices are usually fixed by the competitor. The profit can then disappear very quickly if the efficiency of the process is a few percent lower than the competitor's one or if the quality of the product is slightly lower. It is then easy to understand that a large amount of effort in modern industry is necessary to improve the efficiency and the quality of existing processes and products. Another part of effort is obviously devoted to new products and new technologies but in traditional industries the last one represents usually only 10% of the total effort.

It is necessary to have a deeper understanding of a process to improve it. Most traditional processes were developed as an art and through accumulation of practical experience. Today, such an approach is no longer sufficient and we have to move from an art to a scientific understanding. In industrial processes, matter is transformed in different steps with specialised machines and the flux of matter must be the same as much as possible, through all these machines. The efficiency of these machines must be improved in parallel, or the focus has to be put on the less productive (or the most expensive machine). Industrial equipment is very often extremely expensive and depreciation can contribute from 10 to 30% of the production cost. If it is possible to improve efficiency by 10% for example, the production cost will be reduced by 1 to 3%, which is comparable with the profit of the company.

Calcination of cements and alumina is an example from a very traditional industry: cements as well as hydrated alumina must be calcinated to fire to present the right reactivity with water or to transform hydrated alumina into tabular alumina which could be disolved in hot salt bath in the electrolysis of aluminium operation. Only recently, X-ray and neutron scattering experiments have been done at high temperature to understand all the physical and chemical transformations which have shown to be very complex. From this experimental understanding, it has been possible to reduce substantially the energy consumption and to reduce the dispersion of the properties of the product after calcination, making important savings.

While it is important to have a deeper understanding of processes, it is also necessary to improve products and to have a very good understanding of them. This is very well known in the electronic industry for example where products are created from physical laws which are well understood. It is much less the case in more traditional industries where the knowledge is more empirical. Metallurgy is a very good example of a fantastic improvement through a scientific knowledge in the last twenty years. The thickness of steel for automotive industry has been reduced by 30% with improved mechanical properties, reducing the weight of the cars. This progress has been possible by the work done by physical metallurgy with the help of new characterisation methods. The progress has been done essentially by fundamental research in academic laboratories which was then applied by industry.

An other example of such an improvement is given by the last alloy developed by Pechiney for beverage cans. With the actual understanding of the relations between mechanical properties, micro and macro structures of the alloys and the thermo-mechanical mechanism of formation of these structures, is was possible to make a new alloy with a mechanical strength improved by 18% with the same industrial process (with the same number of steps i.e. without new investments). Such an alloy can reduce the metal content of a can by about 8%. Such a result, like for the steel example, is directly related to the fundamental knowledge accumulated by physical metallurgists.

Hundreds of such examples could be given. However, in many industrial situations, we still don't have the scientific understanding of the process or the product. The list of very exciting subjects could be very long and I will take some examples :

- Grinding of materials : This industrial operation is very common. In many chemical industries materials must be ground to have the best possible reactivity. Two parameters have to be optimised, the energy consumption which is always too high and the size distribution which has to be peaked as well as possible to optimise the reactivity of materials which will be chemically transformed (bauxite -- alumina, iron ore -- iron etc..). Today, we do not have a general theory for material grinding because physicists are probably not aware of this problem which, from a fundamental point of view, is certainly very exciting. We are obliged to make a linear model for a given grinder to try to optimise its operation. A general theory of grinding would probably save hundreds of millions of dollars.
- Rheology of complex granular suspensions : after grinding, it is necessary to handle a material. For concrete, several billion tons have to be handled per year world-wide ! To have a better understanding it is necessary to work in several directions [5] :

  - electrochemistry of interfaces,
  - colloid chemistry,
  - physical aspects of rheology,
  - what to measure and how to measure,
  - reactivity of solid,
  - interaction between mineral and organic materials when one phase is organic additive,
  - chemical synthesis.

  The study is just starting and we are still far from having the needed scientific understanding for a perfect control of cement plaster production or mineral fillers for organic materials etc....
- What is happening in the mechanical contact between a cutting tool and a metal, or between the roll of a rolling mill and the metal sheet or between a dye and the metal during the extrusion process ? Usinor-Sacilor, the French steel industry, is forming with its rolling mills (considering all successive passes) an area equivalent to the area of France. A steel maker is essentially a surface maker and we are still waiting for a good understanding of the process.
- Adhesive bonding : in that field progress have been made with the contribution of De Gennes but still a lot has to be done.
- Emulsions : this field is making progress with the development of researches on mesophases.
- Solidification.
- Multiphase systems.
- Etc.

Here again, the list could be extremely long of subjects which should be interesting for pure physicists and for industry.

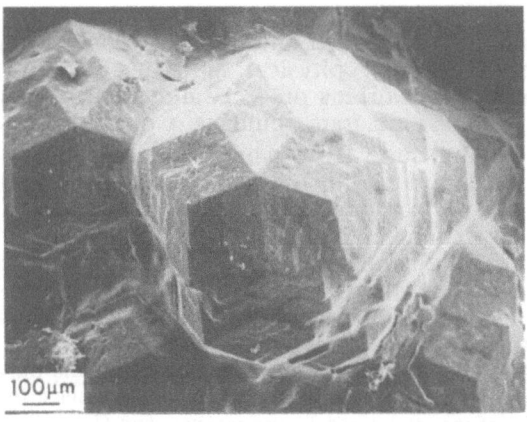

Fig. 3

First observation of a macroscopic quasicrystal exhibiting the fivefold sysmetry (AlLiCu alloy), from B. Dubost et al [6].

IV. HOW TO IMPROVE RELATIONS BETWEEN PHYSICS AND INDUSTRY

Why should such relations be improved ? In this paper arguments were given showing the interest for industry with its need for a deeper understanding of products and processes. It is also interesting for the physics community which could be working on new and very interesting subjects. To be provocative, I found strange to observe how quickly so many physicists were able to shift to high Tc superconductors : such a massive move is probably an indication that these physicists were not working on very interesting subjects and could jump very easily to an other subject. If there is a lack of interesting subjects in the community of physicists, industry can certainly propose a lot of new subjects.

These subjects are really of fundamental nature like thermodynamics. Everyone has to do his own work : applied research must be essentially performed by industrial researchers while fundamental research is mostly the responsibility of academic researchers. It is quite obvious that good interaction is necessary between applied specialists and fundamental specialists. Industrial researchers are not asking academic research to do applied research, they are asking them to perform the best possible fundamental research but to determine their subjects in interaction with the industrial world.

Sometimes also, fundamental research can make important improvement from applied research. This is well know and a recent example in metallurgy will be quoted. A team of physical metallurgists form the Pechiney Corporate laboratory was working on Al Li alloys and realised that a strange composition in the phase diagram could be a quasi-crystalline phase which could crystallise giving quite large monocrystals [6]. The experiment was successful and its was possible for the first time to make quasi-crystals of the order of 1mm size (Fig. 3) and quite quickly much larger. Such crystals have been given to a large number of fundamental physicists who were then the first to be able to make neutron or X-ray scattering, to study properties of such new phases. AlLi alloys are not developing industrially but this research had an unexpected output, the study of a new organization of matter (icosahedral symmetry).

The question arises as to how to improve the relations between physics and industry ? In the field of analytical methods, it works usually quite well. However, as said before, it would be necessary for the academic world to re-emphasise the importance of instrumental work in the laboratory. Some academic engineering departments should make still more measurements (applied research) for small companies which cannot afford the equipment and are not rich enough to hire good specialists of these techniques. For very large facilities, percolation is usually slower for several reasons : cost of the beamtime, lack of training from the industrial side, lack of scientific support from the facility side ; these problems could be solved by more interactions on each side.

For heavy physics, it would be very interesting to create relations either with large companies specialised in engineering or with engineering of large companies. As said previously, they could learn a lot from each other and the first could train students who could be hired by the second. It is not really physics but it is technology which is necessary to perform physics as well as to improve production. It should be normal for a lot of young high energy physicists who specialised in different technical fields to join industry after the PhD. That is not yet the case.

But certainly the best way to improve relations is the exchange of people. In our corporate laboratory we always have some professors from many institutes spending their sabbatical year : they learn about our problems and afterwards work on fundamental aspects in their own laboratories. We also do have a lot of consultants spending part of their time with our company. That is a valuable contribution for many companies and it is very efficient also for the training of PhD who could be hired by the company. The reverse is less common : not enough industrial researchers are spending part of their time in academic laboratories. A large part of the responsibility is from the industrial side which does not appreciate enough the interest of such an exchange and are not given the needed time ; academic side is also partly responsible because they are often not able to propose interesting positions.

CONCLUSION :

We have shown in this paper that physics has contributed a lot to wealth and protection by inventing a lot of new technical systems which have changed our life.
Since the sixties and mostly the eighties, there is a clear change and industry is asking physicists for heir help to have a better understanding of their processes and their products, in order to improve them. In parallel, it

appears that questions raised by industry can be new subjects for fundamental physics.

In science, subjects have always emerged from science itself but also from the daily experience in the external world; probably physicists have far too long neglected intersesting subjects the second source. The interaction between physics and industry is usually on-going on existing or new analytical methods. For heavy physics where the engineering part is extremely large, contacts with industrial engineering would be useful to improve on both sides.

As D. Kleppner said in a round table published recently [7] "what is missing today is a sense of the future". Each generation must invent its future. Research must be part of this process. The future is the same for everybody. It has then to be invented with people strongly interacting with each other. A good interaction between physics and industry will contribute at its level to the invention of our future.

References :

1-   P. Cannon - Round Table "Physics Research in Industry" - Physics Today - Feb. 88 - p. 54

2-   From PJ. Webster - Steel Times (June 1990).

3-   JC. Conesa, P. Esteban, H. Dexpert, D. Bazin - Spectrosocpic Characterization of Heterogeneous Catalysts - Editor : Elsevier 57, A225 (1990).

4-   SR. Amendolia - Nucl. Inst. Methods A 263 - 155 (1988)

5-   I want to thanks J. Lukasik, scientific director of Lafarge Coppée for this example.

6-   B. Dubost, JM. Lang, M. Tanaka, P. Sainfort and M. Audier - Nature 324 - 48 (6 nov. 1986).

7-   Round table : Science under Stress - Physics Today - Feb. 92 - page 38

THE ROLE OF UNIVERSITIES IN THE DEVELOPMENT OF JAPANESE SCIENCE

Akito Arima
Hosei University*
3-7-2 Kajino-cho
Koganei-shi, Tokyo 184, Japan

Abstract   Japanese universities are distinguished by long-standing strengths in engineering and applied technology, with much less emphasis being given to natural sciences and basic research. By tracing the historical development of the Japanese university system, it is possible to identify factors that have shaped this bias towards engineering. Such an approach is used in this paper to explain the present characteristics of the Japanese university system, together with factors affecting the changing relationship between academic and industrial research. While a strong engineering base will continue to be a vital element in supporting Japan's future economic development through the supply of large numbers of high-quality engineers and technologists, there is also an argument for expanding university science.   Now that Japan has reached a position at the frontier of international best practice in many fields of technological development, there is an ever-greater need for Japan to play a larger role in the international development of basic science. This will help to stimulate original thinking and also create a potential for supporting the future development of new technologies. In consequence, this paper proposes that Japanese science should be expanded in harmony with a continued commitment to developing existing strengths in technology.

Introduction

The present status of Japanese science has been shaped by the development of Japan's university system and its historical bias towards engineering. This paper therefore begins by reviewing the history of Japan's university system to explain its current emphasis on applied technology and engineering, rather than science. The relatively low priority given to university science has also been reflected in the industrial sector, where firms have traditionally been more concerned with development-oriented research. However, there are signs that this is changing, thereby paving the way for increased cooperation between industry and academia. In conclusion, the paper agues that basic science and applied science both form essential elements in the processes by which knowledge is created for the greater benefit of society. That is to say, science and technology should be seen as partners in a harmonious relationship. While Japan has established strengths in technology, there is also a need for it to play a larger role within the international scientific community.

A Short History of Japanese Universities

Japan's university system can be traced from the foundation of the University of Tokyo in 1877. This move into higher education was one of many changes taking place in

* Present address
Institute of Physical Chemical Research
2-1 Hirosawa
Wako-shi, Saitama 351-01, Japan

Japan at that time. In 1868, the Meiji Restoration marked the establishment of a non-feudal central government in Tokyo, which was committed to improving the education system as part of a broader package policies to develop the Japanese economy in the wake of the country's long period of seclusion from the outside world. This led to significant increases in the adoption of foreign organizational systems and technologies. In this respect, the development of higher education in Japan was initially shaped by a number of overseas influences.

The University of Tokyo originally had four faculties: law, science, medicine and letters. Within the Science Faculty there were departments of physics, mathematics, astronomy, chemistry, biology and geology, as well as three more applied disciplines: mining, civil engineering and mechanical engineering. Like many leading universities in Europe, there was no separate Faculty of Engineering. In Europe, applied science and technology were taught mainly in colleges and polytechnics -- leading to the development of some outstanding institutions such as France's Ecole Politechnique and the München Technische Hochschule in Germany (which became the München Technische Universität at the end of the Second World War). While many European universities developed extremely strong science faculties, the development of engineering was less pronounced. Even today, many prestigious European universities have relatively weak engineering faculties. Moreover, shades of this European tradition can also be seen in the structure of US universities. For example, while Harvard University has a strong department of physics, there is no independent department of engineering -- although considerable strengths in this area have been established at the university's near neighbor, MIT.

By contrast with the "European tradition", the development of Japan's University system took a rather different course. As part of an effort to promote technical education in Japan, the Ministry of Industries hired a Scottish railway engineer -- Dr. H. Dyer -- to advise on the setting up of a college for engineering. These plans led to the foundation of the Tokyo College of Engineering in 1873, which was headed by Dyer and had a number of professors recruited from Britain. In 1886, the college was absorbed by the University of Tokyo -- which then changed its name to become the Imperial University. The formation of the University's Faculty of Engineering followed as a result of separating the existing Department of Engineering from the Faculty of Science and merging it with the activities of the Engineering College.

In 1891, the Imperial University absorbed the Tokyo College of Agriculture, which was then restructured as the University's Faculty of Agriculture. Much later, other faculties were added such as Economics, General Arts and Sciences, Education and Pharmacology. However, a key point to note is that, by the early 1890s, the University had a strong base in practical disciplines such as engineering and agriculture, thereby signaling a departure from the European tradition.

Japan's second university was established in 1897 at Kyoto and was known as the Imperial University of Kyoto (whereupon the Imperial University in Tokyo changed its name to the Imperial University of Tokyo -- although, after the Second world War, both universities dropped the "Imperial" appellation from their titles). The new university in Kyoto shared the same structure as its counterpart in Tokyo, with a strong emphasis on engineering and agriculture. Aspects of this structure were subsequently adopted by other major national universities such as Kyoto, Hokkaido, Tohoku, Nagoya, Osaka and Kyushu -- thereby creating a tradition which became increasingly common as the university system expanded.

Characteristics of Japan's University System

The expansion of the Japanese university system is shown in figure 1. By 1950, some 50 universities and colleges had been established. After the Second World War, there was a rapid increase in the number of universities. This was initiated under the American Occupation and continued after the Occupation Forces withdrew. There are now 534

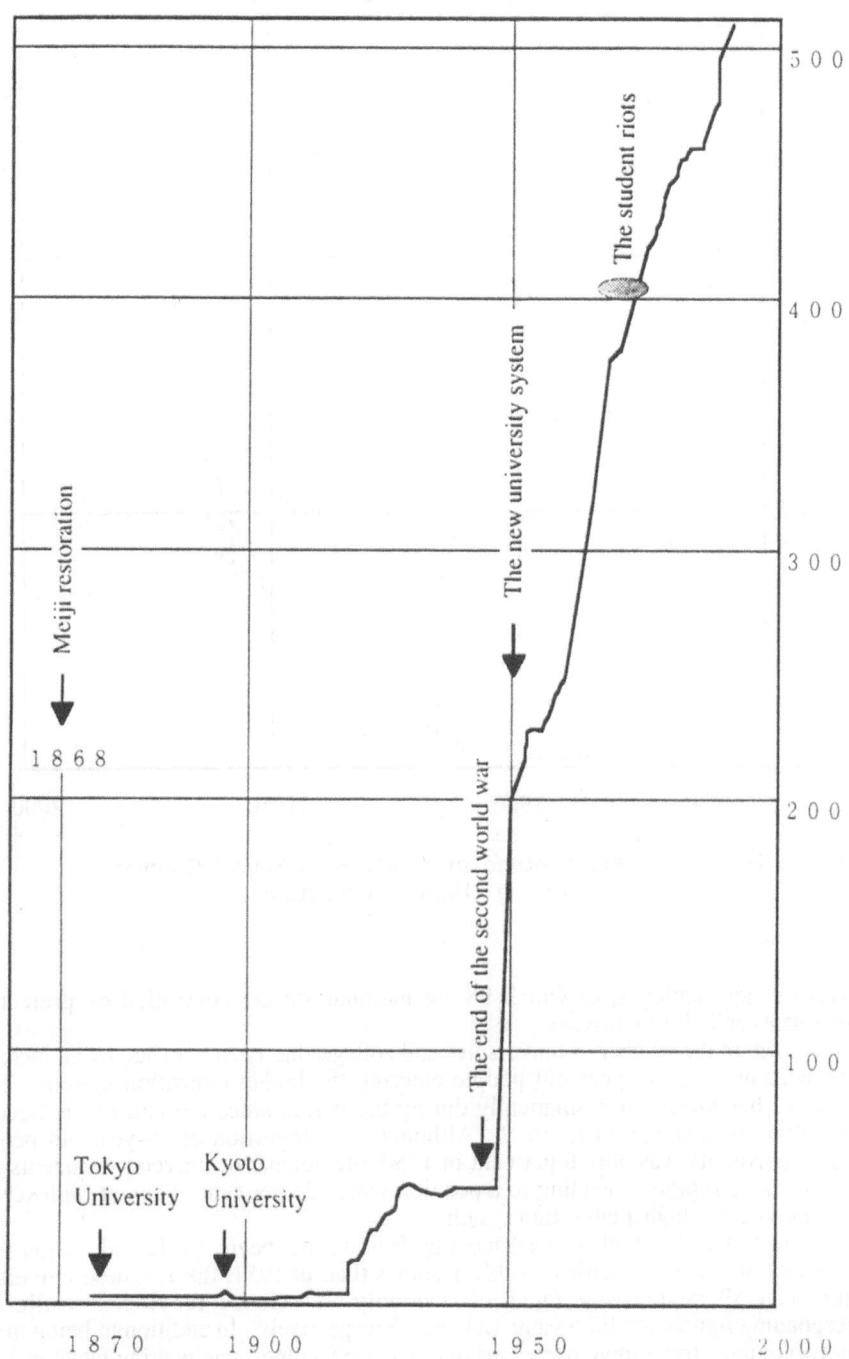

Fig 1   Number of Japanese Universities

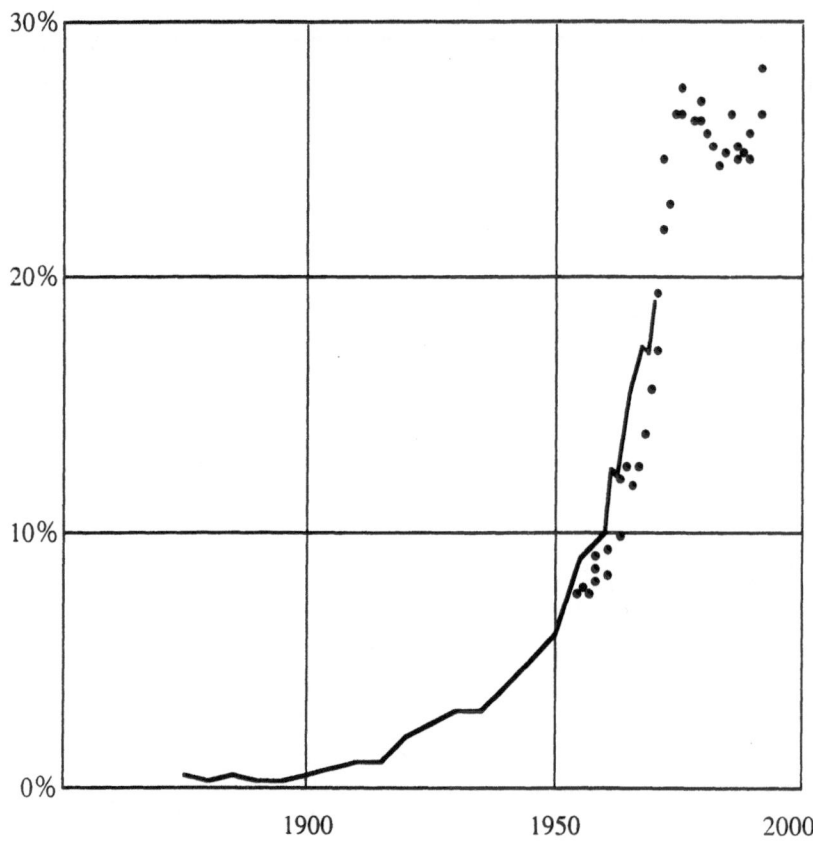

Fig 2   Per centage of Senior High School Graduates
              Entering   Higher   Education

universities and colleges, of which 98 are national, 46 are controlled by prefecture
governments and 390 are private.

Growth in the number of universities and colleges has been matched by an increase
in the percentage of 18-year-old people entering the higher education system. This
percentage has increased dramatically during the period since the end of the Second
World War, as is shown in figure 2. Although the proportion of 18-year-old people
going to university was only 8 per cent in 1950, the number of university students has
since increased rapidly -- leading to a position where 25 per cent of senior high school
graduates enter the higher education system.

Historically, the number of engineering faculties has been significantly larger than
the number of science faculties. Table 1 shows that, in 1951, the Japanese university
system had 59 engineering faculties and only 21 science faculties, while the
corresponding figures for 1985 were 124 and 45 respectively. In addition to being in the
order two or three times more numerous than science faculties, engineering faculties have
also tended to be rather larger.

Table 2 shows that, in 1989, engineering faculties in Japanese universities were
attended by some 378,405 students, while the corresponding figure for science faculties

		NATIONAL	PROVINCIAL	PRIVATE	SUM
1951	E	32	5	22	59
	S	14	1	6	21
1965	E	40	5	49	94
	S	19	3	11	33
1985	E	51	5	68	124
	S	30	3	12	45

**Table 1   NUMBER OF FACULTIES OF ENGINEERING
AND SCIENCE**

was only 63,997 students. This ratio of nearly 6-to-1 in favor of engineering distinguishes Japan from other advanced industrial countries and has made a major contribution to its rapid economic development, particularly during the period since the Second War.

Some international comparisons, which illustrate the extent of the Japanese higher education system's emphasis on engineering, are provided in Tables 3 and 4. In both tables, Japan's production of graduates with bachelor and Ph.D. degrees in science. engineering and agriculture contrast sharply with comparable data for the USA. the United Kingdom and the former West Germany. While Table 3 expresses the data in terms of student numbers, Table 4 shows the ratio of science graduates to both engineering and agriculture graduates.

At the bachelor level, Japan's relatively low number of science graduates -- compared to engineering and agriculture -- is very different from the other three countries which all exhibit strong science sectors. In the case of the United Kingdom and West Germany, the number of bachelor degrees in science awarded in 1988 exceeded the total number of engineering graduates by a wide margin. While the USA produced more bachelor degrees in engineering than science, the disparity is far smaller than in the case of Japan.

Tables 3 and 4 also show that Japan's emphasis on engineering relative to science is evident at the Ph.D. level -- although it should be noted that Japan produces comparatively few Ph.D.'s. (Japanese industry generally has a strong preference for in-

(Unit:Person)

	1985	1989
Total	1,734,392	1,929,137
Humanities	246,850	290,387
Social sciences	671,001	759,636
Natural sciences	59,678	63,997
Engineering	342,590	378,405
Agriculture	60,068	64,975
Health	117,809	117,712
Home economics	32,185	35,794
Education & teacher training	135,227	139,565
Arts	44,890	47,005
Others	23,094	31,661

Note: Others include department of marchant marine.

Source: Ministry of Education, "Report of Basic Survey on Schools", various educations.

Table 2   **Number of Students in Colleges and Universities by category of department (3)**

house training and, rather than wait for a student to complete a doctorate, the majority of Japanese firms would prefer to recruit graduates at the bachelor level and train these individuals in accordance with company procedures.) The position is rather different in the other three countries. For example, Table 3 indicates that the United Kingdom (which has less than half the population of Japan) produced many more Ph.D.'s, with a heavy emphasis on science rather than engineering. Thus, while the United Kingdom's production of science Ph.D.'s was more eight times larger than the comparable figure for Japan, the number of engineering Ph.D.'s was only about two-and-half times greater.

Bachelors/year	Science	Engineering	Agriculture
Japan(1990)	14,217	86,115	14,854
U.S.(1988)	68,520	126,341	13,488
U.K.(1988)	21,900	15,200	1,500
W.G.(1988)	13,106	11,554	2,920
PhD/year			
Japan(1989)	876	1,774	734
U.S.(1988)	8,267	5,163	1,184
U.K.(1988)	7,200	4,500	900
W.G.(1988)	4,343	1,424	989

Table 3   Number of bachelors and PhD/year

In the past, the majority of Japan's engineering students came from national universities, while private universities tended to be stronger in social science and humanities. However, in recent decades, the number of provincial and private universities with faculties of engineering has increased -- which has served to further strengthen the higher education system's ability to provide industry with large numbers of highly-trained technical graduates.

At present, the greater part of funding for university research comes from the Ministry of Education, Science and Culture (Monbusho), which offers two forms of support. One is awarded on the basis of a head-count of the number of staff, while the other -- which is known as the "Grant in Aid of Science" -- is allocated on a selective basis in response to applications. While the size of the "Grant in Aid of Science" has doubled during the period since 1984, there is still a need to further increase this area of expenditure. Although Japan's R&D -- expressed as a percentage of GNP -- is the

Bachelors/year	Science	Engineering	Agriculture
Japan(1990)	1	6,1	1,0
U.S.(1988)	1	1,8	0,2
U.K.(1988)	1	0,69	0,07
W.G.(1988)		0,88	0,22
PhD/year			
Japan(1989)	1	2,0	0,84
U.S.(1988)	1	0,62	0,14
U.K.(1988)	1	0,63	0,13
W.G.(1988)	1	0,33	0,23

Table 4   Ratios of bachelors and PhD.

highest in the world, the majority of this expenditure is made by companies. Inevitably the bulk of this industrial research is directed towards application-oriented development projects. While there has been an increased tendency for industry to stress the long-term importance of basic research, many such initiatives tend to put weight on the basics of applied research rather than basic science.

Cooperation between Japanese Universities and Industry

Examples of cooperation between industry and universities have tended to be most common in the field of engineering, although there are some notable cases of scientific collaborations. One classic example centered on the development of iron for magnets carried out just before the Second World War by physicists like Kotaro Honda, who worked at Tohoku University in Sendai. There are also many post-war examples of

Source	IN 10^8 Y
Ministry of Education, Budget depending on number of professors	1,640
Grant in Aid of Science	824
Industry	522

Table 5   TOTAL BUDGETS OF RESEARCH AND EDUCATION OF NATIONAL UNIVERSITIES (SALARY IS NOT INCLUDED)

collaboration in semiconductor components, new materials super computers and so on. However, it is difficult to deny that cooperation between industry and academia was seriously weakened by student unrest which spread throughout Japan from 1968 to 1975. At that time, militant students were critical of industrial involvement in the university sector. Although the influence of this type of thinking subsequently diminished, the idea that research done at national universities should be pure and therefore supported by government retained some considerable currency until around 1985. Researchers at national universities were often reluctant to be seen to be collaborating with industry. This picture has since changed. In recent years, industry has become an increasingly more important source of funding for universities and is generally regarded to be a welcome supplement to government funding.

The size of industry's contribution to the expenditure of national universities is shown in Table 5. This illustrates that -- excluding staff salaries -- industry's contribution to research and education at national universities has grown to the point where it is approaching 20 per cent of the expenditure provided by Monbusho. While industrial money should not be seen as a substitute for government funding, it can provide a useful mechanism for reducing barriers between industry and academia. By securing further increases in both government and industrial funding for science, it should be possible to ensure that Japan is able to play a larger role in the international development of basic research.

Future Directions for Japanese Science

The high priority give to engineering within the development of Japan's higher education system has been a major factor in supporting the country's economic development. In the past, success in engineering has often distracted attention from the relatively slow progress that has been made in developing Japan's scientific research base. However, in the future there is likely to be a greater need for the type of knowledge that can be created through basic research and original thinking. In recent years, Japan has attracted criticism both at home and abroad for its relatively limited contribution to basic science.

At the same time, Japan can no longer look to other countries in order to form its strategies for technological development -- it has to generate creativity and make a full contribution to the development of basic science.

In some respects, Japan's current industrial strength and the relative weakness of its commitment to basic research might be analogous to the position of the USA in the early-1940s. During the first 40 or so years of the twentieth century, there was a rapid expansion in the USA's industrial development and the establishment of a number of "science-related" industries. At that time, much of the knowledge base which supported these science-related industries originated from European research into basic science and technology. However, the subsequent development of American science proved to be spectacular and the USA emerged as a global leader in the pursuit of basic research.

Now that Japan has become established as a leading advanced industrial economy, there is a growing need to complement expertise in technology and engineering with a larger contribution to the development of basic science and, in particular, physics. (On the other hand, in the case of many European universities and -- to a lesser extent -- American universities, it could be argued that there is a need to match strengths in science with increased attention to the development of engineering faculties.)

Education is a vital resource for underpinning the development of advanced industrial societies. In this respect, the improvement of both technological and scientific education are important objectives for modern education systems. Technology and science represent different aspects of broader knowledge creating systems. It is therefore important to ensure that there is an appropriate balance between applied and basic research -- we need them both!

A NEW PARADIGM OF SCIENCE POLICY AND TECHNOLOGY MANAGEMENT FOR THE COMING CENTURY

Y. Takeda
Hitachi, Ltd.
1-5-1 Marunouchi, Chiyoda-ku, Tokyo 100, Japan

Abstract. In this century we have encountered unprecedented situation of a population explosion and expansion of economic activities with massive consumption, and now we are facing world-wide problems resulting from these phenomena. To create prosperous society by releaving strains caused by the pehnomena, we must have a new paradigm for industry and technology development based on a new sense of values, i.e., life of nature and human beings must always be taken into consideration. The new technologies based on this new paradigm can only sprout on new scientific principles which will stimulate enthusiasm to the younger generation of the next century. Synergistic endeavors to overcome any possible obstacles will be needed to create new technological innovations.

INTRODUCTION

I would like to express my sincere thanks to the organizing committee of IUPAP Academic Session for inviting me to this conference. I am very honored to be given a chance to make a presentation in this highly respected international meeting and at this wonderful place.

Today, I would like to talk about "A new paradigm of science policy and technology management towards the 21st century". Before stating my expectations for the next century, let me reflect on the history of human beings very briefly. Human beings have experienced at least two major revolutions in the way of production. The first is the transition from the hunting-gathering society to the

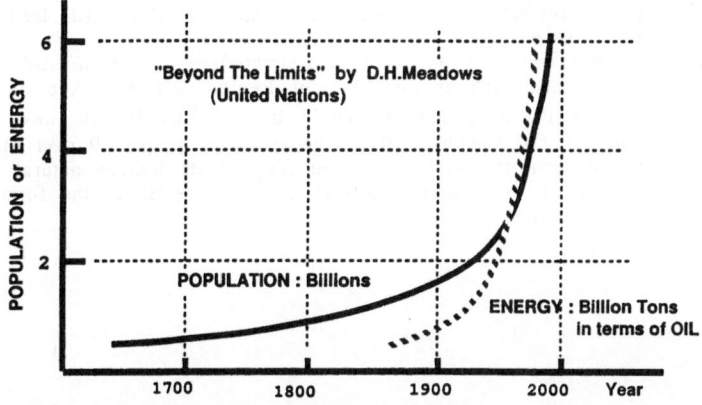

Fig. 1 WORLD POPULATION & ENERGY CONSUMPTION

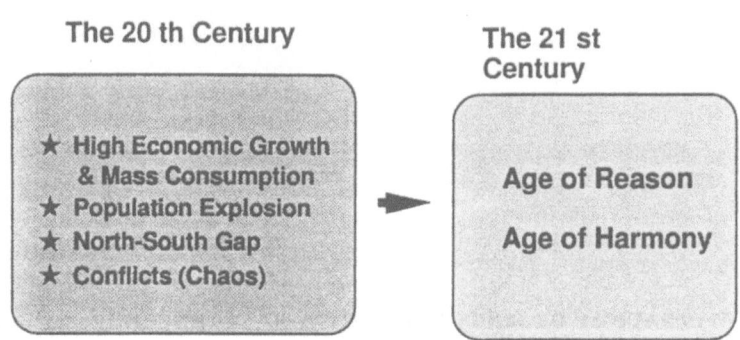

Fig.2 EXPECTATIONS FOR THE COMING CENTURY

agricultural society, which occurred about 5000 years ago in regions along the Nile river, the Tigris-Euphrates river, the Indus river and the Hwang Ho river. In Japan this transition happened about 2200 years ago, in the Jomon period to Yayoi period. Since the productivity of foods increased about ten times in those days, the population increased about 10 times accordingly.

The second is the transition from the agricultural society to the industrial society, known as Industrial Revolution which started in the 18th Century in England. Due to the Industrial Revolution, population increased sharply by 3 - 5 times in countries which underwent industrialization. In Japan the industrialization happened after the Meiji Restoration in 1868 and the population increased about 3 times in the following 100 years and levelled off in the last few decades. Now at the end of the 20th Century the massive consumption of natural resources is still continuing in advanced countries, and on the other hand, the population is soaring in developing and under-developed countries(Fig.1).

I imagine that future historians will call the 20th Century, "the age of extreme growth and consumption, the age of population explosion, or even the age of chaos" (Fig.2). It is true that during the last century, the World's economy overall has grown about ten times. However, at the same time, this growth is associated with the huge consumption of natural resources, including coal, oil and minerals. Consequently, we have encountered unprecedented global problems such as the warming-up of the earth and the destruction of the ozone layer that were never even thought of by people before the 19th Century. Along with the development of a world economy, expansion of the north-south economic gap and tragic regional conflicts are still continuing. Unless countermeasures against these situations are undertaken globally, the outlook of human beings in the 21st century will not be bright. Some people predict that the population explosion alone will lead to a geo-catastrophe in the 21st century.

I hope the forthcoming century will be characterized as a wonderful age and be remembered by future historians as the "Age of Reason" and the "Age of Harmony". I hope the society in the 21st century should be the "**Intelligent and Ecological Society**". That is, the 21st Century will be the age when people all over the world will cooperate to conserve the global environment and its limited natural resources. To realize this goal we have to prepare beforehand to cooperate in the formation of a global-scale "sense of values".

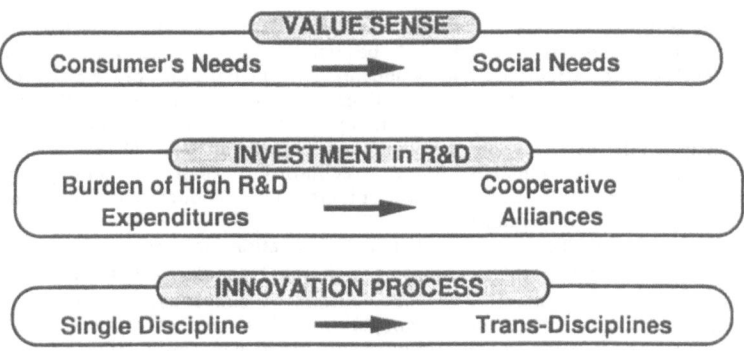

Fig.3 CHANGES in SOCIAL CURRENT

Therefore, I propose here today that the guiding principle of the science policy and the technology management for the 21st Century, should be "the contribution to human progress with a deep consideration for the stability and harmony of the earth", rather than only growth through the realization of economic benefits to human beings.

SYMPTOMS OF CHANGES IN SOCIAL CURRENT

The symptoms of this big change are observed as several flows in the present society already moving towards the next century. I would like to point out several such promising symptoms of change, from among the current activities of private corporations, especially activities related to the technology management for the benefit of humankind (Fig.3).

The **first** is the change of "value sense" underlying industrial activities. It is quite obvious that a major shift in business policy is now being undertaken from the sole consideration of consumers' needs to those much wider social needs. Accordingly, the important areas of R&D in industries, for example in Hitachi, are shifting to :

(1) Energy, environment, and the recycling of resources
(2) Social infrastructure (such as water supply, drainage systems, transportation systems, and information network systems)
(3) Systems related to health, medicare and human safety
(4) Electronics and software for the realization of an advanced information network society

The **second** is the change in the relationship between investment in R&D and its return. For the past 15 years, since the oil crises of the 1970's, high-technology industries, including Hitachi, have steadily increased their R&D investment. At the same time, during these years, the economic return from the business has also increased in a seemingly proportionate way. Thus, competition in R&D has started and has become over-heated, resulting in a shortening of product life-cycles. Eventually returns from the R&D investment gradually decreased. But it is impossible to cease the investment in R&D in order to stay competitive in business. Thus, the so-called "High-Tech Syndrome" has become a rather over-used phrase in advanced countries.

Some reflections on this situation have come from among industry leaders. The importance of a "*cooperative mind* " became much more advocated. In other words, the necessity for effective and concentrated utilization of R&D resources has increased. Furthermore, cooperative alliances between companies all over the world are now emerging. For example, at Hitachi, we are now promoting cooperation with Texas Instruments company in semiconductors such as 256 M bit DRAM and with General Electric in electric power technology such as gas turbine, power distribution systems and atomic power plants. This cooperative R&D trend will be accelerated more in the future.

Third, I would like to stress that the innovation process, itself, is also changing. It is an image of the past, where some individual genius makes an invention based on one scientific principle or one single technology, in a single cultural environment, without any help or cooperation from others. I believe this kind of "Lone-Ranger" innovation could only have happened until the beginning of the 20th Century.

Today is the age when technology innovations can only be realized through the cooperation of many researchers or engineers, utilizing many scientific principles and feasible technologies. Also, information related to such technology innovation is communicated all over the world and across boundaries of different geography and different cultures. That is, we now are in the age when "many-many" people on the earth will participate in realizing the innovation. However, attention must still be paid to the fact that technology innovation is not born from the mere summing up of linear activities, but rather, is brought about through an interactive "**synergy effect**" of people working together.

During my early days in Hitachi in 1960's and early 1970's, I was a researcher in the optoelectronics area. Therefore, I have always paid attention to the research and development process of laser technology, and the process by which its development made a contribution to human society through its applied technologies, such as optical-communication, compact disks, laser printers, and laser surgery. I think this is a pioneering example of the technology innovation process through the synergetic interactions among people in many kinds of disciplines, technologies and social needs. In this field, as you know, many researchers from the United States,

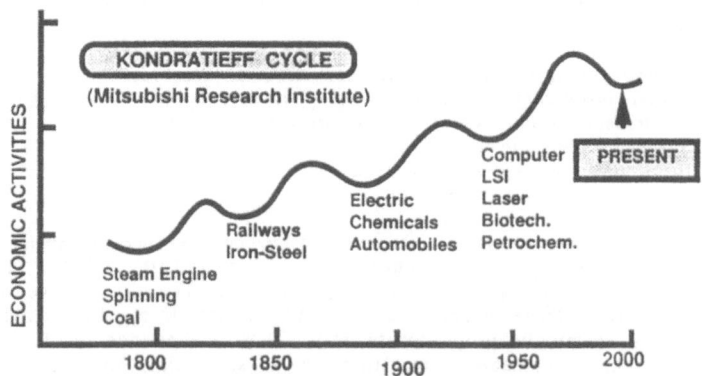

Fig. 4 CORRELATION BETWEEN ECONOMIC ACTIVITIES
AND TECHNOLOGY REVOLUTION (Conceptual)

Europe, Russia, and Asia including Japan contributed to the realization of this technology innovation.

ECONOMIC ACTIVITIES AND REVOLUTIONARY INNOVATIONS

When we look back the history of the industrialization in the past 200 years, we observe several waves of economic activities and revolutionary innovations(Fig.4). As is exemplified in "Kondratieff Cycle", we recognize 4 big waves in the economic activities and technological innovations approximately in every 50 years. The first one is, of course, the industrial revolution started in England around 1800, which may be characterized by a steam engine, spinning and coal-mining industries. We can see the second wave around 1850 and the third around 1900. The last and 4th wave started, I suppose, around 1950 and economic activities were driven by technologies such as computer, LSI, laser, biotechnology and petrochemicals. I think advanced countries enjoyed the economic booms after World War II as a whole. We can learn several lessons from the history(Fig.5). *First*, sprouts of technological revolution appear at the bottom of economic activities. *Second*, progress of technological innovations has a positive correlation with the economic activities. *Third*, the economic activities start to decline at the saturation of technological innovations.

At present the world economy is in recession with an exception of South-East Asia. I might say that technological innovations of the conventional industries, which have been the driving force of economic activities in the last 50 years, have reached a saturation. Now is the time that we can expect revolutionary technologies to emerge, which can induce new industries.

TECHNOLOGICAL INNOVATIONS AND
ECONOMIC ACTIVITIES

1. SPROUTS OF TECHNOLOGICAL REVOLUTION

APPEAR AT THE BOTTOM OF ECONOMIC ACTIVITIES.

2. PROGRESS OF TECHNOLOGICAL INNOVATIONS HAS

POSITIVE CORRELATION WITH ECONOMIC ACTIVITIES.

3. ECONOMIC ACTIVITIES DECLINE AT THE SATURATION OF

TECHNOLOGICAL INNOVATIONS.

Fig.5 LESSONS FROM THE PAST HISTORY of 200 YEARS

SOCIETY IN THE 21st CENTURY

Society in the 21st Century, or more specifically in the first half of the 21st Century will be quite different from that of today(Fig6). We observe a lot of changes

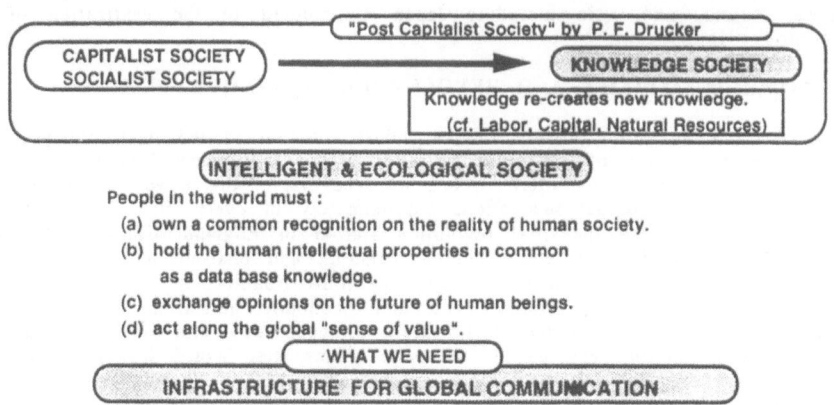

Fig.6 A NEW SOCIETY in 21st CENTURY

currently happening in our society, and some of them will definitely become main streams in the future society. There are several key phrases for the future society : "Post-Industrial Society", "Advanced Information Society", and "Information and Service Society". Recently, Peter Drucker published an interesting book entitled "POST-CAPITALIST SOCIETY'. He foresees that "KNOWLEDGE SOCIETY" will come out after the capitalist society which is already quite different from that defined by Karl Marx in the late 19th century. In the knowledge society *Knowledge* plays a major role in social activities, including operation of private and public organizations. In a traditional industry, labor, capital and land (i.e. natural resources) are key factors of production. [I quote from "Post-Capitalist Society".] Knowledge is now fast becoming the sole factor of production, sidelining both capital and labor. From now on, what matters is the productivity of non-manual workers. And that requires applying knowledge to knowledge. In the coming society, knowledge is the means to obtain social and economic results. Knowledge is the only meaningful resource today already.

I have stated my expectations for the next century, i.e., it will be the "Age of Reason". Knowledge originates from reason. The knowledge society is based on reason. The knowledge society should be the "Intelligent and Ecological Society". In order to apply knowledge for happiness of global society, people in the world must;

(a) own a common recognition on the reality of human society.
(b) hold human intellectual properties in common as a data base knowledge.
(c) exchange opinions on the future of human beings.
(d) act along the global "sense of value".

Although the above concepts are not easily achievable, these are minimum requirements to solve the pending problems, such as the global environment, north-south gap, population explosion and regional conflicts. I suppose that knowledge society will be quite different from a so-called "Information Society" in the contemporary meaning. In order to achieve our goal, it is prerequisite to construct infrastructure of global information network for all people to communicate each other freely and instantaneously, and for all people to hold

```
★ INFRASTRUCTURE FOR            ★ ENERGY & RESOURCES
  GLOBAL COMMUNICATION

★ ELIMINATION OF                ★ EARLY WARNING SYSTEMS
  LANGUAGE BARRIERS               FOR NATURAL DISASTERS

★ GLOBAL ENVIRONMENT            ★ MEDICAL & HEALTH CARE
```

Fig.7 CHALLENGING INNOVATION FIELDS TOWARDS 21st CENTURY

human knowledge in common. Since there exist so many languages, say more than several hundreds in the world, it is also very important to develop an automatic translation system to overcome language barriers.

I summarize here the challenging innovation fields towards the "Intelligent and Ecological Society" (Fig.7) : (a) Infrastructure of global communication, (b) Elimination of language barrier, (c) Global environment, (d) Energy and resources, (e) Early warning systems for natural disasters, and (f) Medical and health care. Although most of these subjects have already been undertaken in private and governmental organizations in advanced countries, however, I do not think these subjects will be handled by the present high technologies and their extensions alone. In the past,
development of the semiconductor and laser technologies was based on Solid State Physics and Quantum Mechanics. We know that advancement of science precedes revolutionary innovations. Today, different types of basic sciences are desired. However, new sciences such as Ecological Science, Geo Science, New Energy Science, Medicare Science and Life Science are not yet established to induce revolutionary innovations requested. Although people are eagerly waiting for new technologies to come out soon, even in advanced countries a basis for *New High Technologies* is not prepared. Participation and contribution from many scientists and engineers from all fields, of course including physicists, are required for the realization of the "Intelligent and Ecological Society".

SOCIAL SYSTEMS FOR REVOLUTIONARY INNOVATIONS

In the past fifty years, scientific achievements from fundamental researches have been converted into technologies, and then commercialized mainly by private corporations in the capitalist society. Marketing new products to society has been a main source of company's prosperity. Profits from the sales are invested into new innovations(Fig8). Today so-called high-tech companies invest as much as 10% of the total sales into R&D.

To realize "Intelligent and Ecological Society" in the 21st Century, a new paradigm of science policy and technology management will be necessary. Since technologies to solve the pending problems such as global environment and the global communication infrastructure, are so widespread and long-ranged, huge amounts of R&D expenditures and human resources will be required. Furthermore, the expenditure is more than private corporations alone can bear. Therefore, we

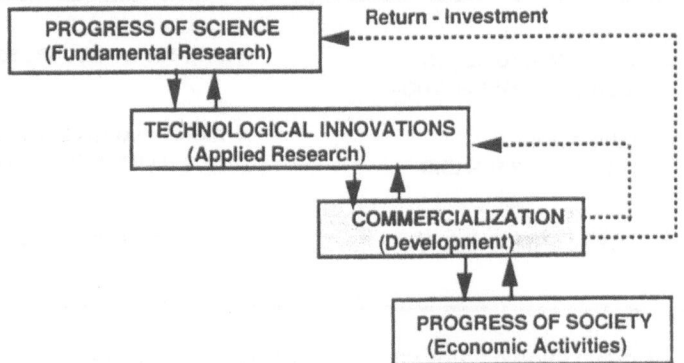

Fig. 8 PROCESS OF COMMERCIALIZATION OF NEW TECHNOLOGIES

need public sectors including academic sector to fund and invest science and technology to induce revolutionary innovations in these fields(Fig.9). Also, a flexible technology management system to further enhance cooperation between Academia, Governmental and Private sectors must be established, i.e., we have to overcome barriers existed between them in the past.

It is requested especially for advanced countries to invest continuously in these fields to achieve our great goals. This is considered to be the utmost and urgent responsibility of advanced countries to the world. Considering the correlation between CO_2 Emission and Gross Domestic Product, it is clear that the industrialized countries must take an initiative to suppress their CO_2 emissions(Fig.10). National consensus, however, must be formed to establish the funding system and the national budget must reflect public opinions.

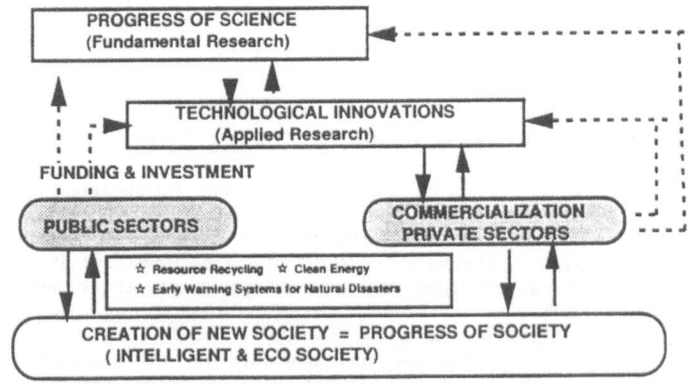

Fig. 9 NEW FUNDING SYSTEM FOR SOCIETY ORIENTED TECHNOLOGIES

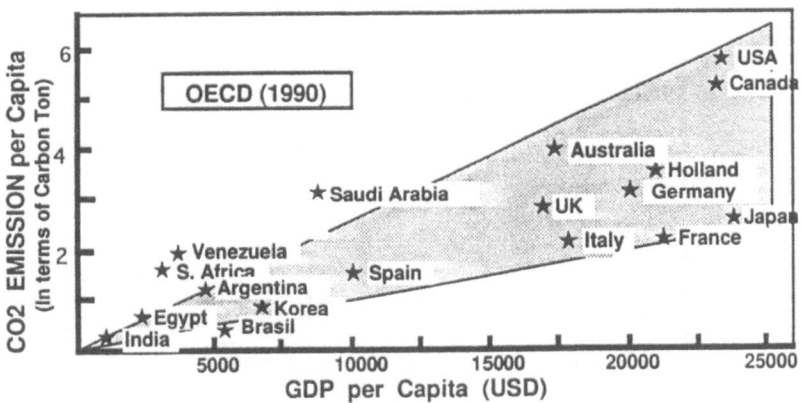

Fig. 10   CO2 EMISSION vs GDP Per Capita

In the present world there are countries at various stages of economic development, i.e., advanced, developing, and under-developed countries. In advanced countries creation of new industries, which suit to the "Intelligent and Ecological Society", should be the first priority. In developing countries advancement of current industries should be the first priority, and in under-developed countries build-up of basic industries should be the first priority. Supports are requested in transferring technologies from advanced countries to developing and under-developed countries(Fig.11). We expect that diffusion of innovations and technologies over the world will be much faster in the 21st Century.

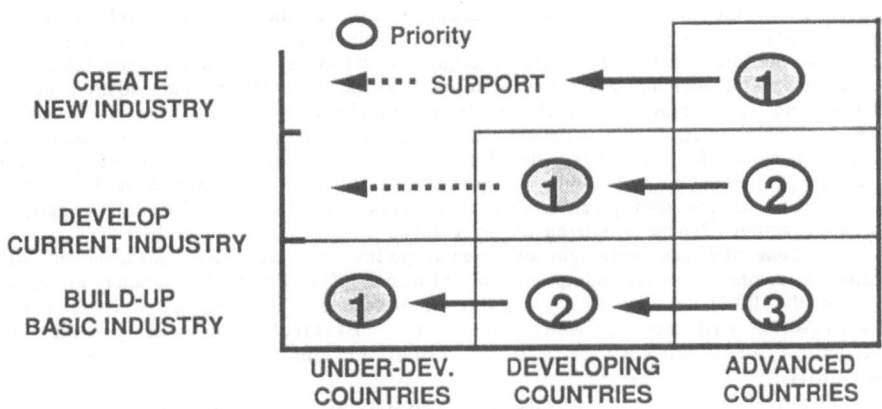

Fig. 11   GLOBAL DIFFUSION OF TECHNOLOGIES
ROLE OF ADVANCED COUNTRIES

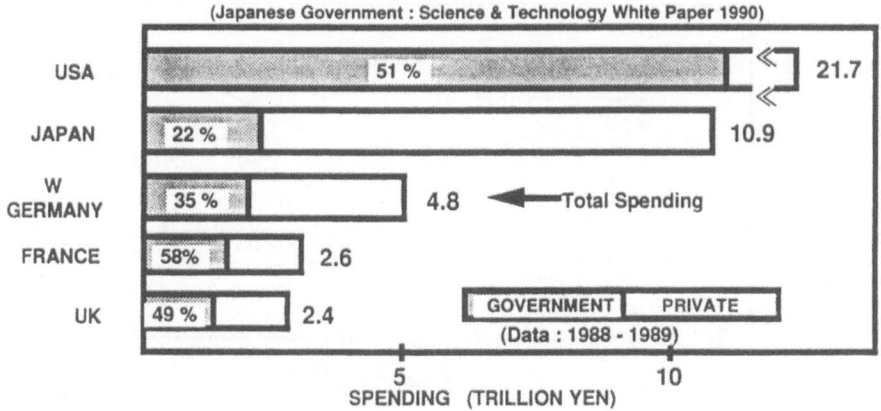

FIG. 12 COMPARISON OF NATIONAL R&D EXPENDITURES
(Ratio of Governmental Spending)

ROLE OF ADVANCED COUNTRIES --- JAPAN

As I stated previously, to create a new society in a desirable form in the coming century, understanding and support of citizens, especially in advanced countries, are necessary. In most advanced countries with an exception of Japan, government spending for R&D constitutes about 30 - 50 % of the whole national R&D expenditure (Fig.12). In the past, about one half of the governmental R&D was used in developing military technologies. In the age of EAST-WEST confrontation it was inevitable, I must admit regrettably. But now military-oriented R&D is no longer required as in the past. If the total budget is kept at the previous level, the saved money should be spent to realize a new global society with reason and harmony.

On the other hand, the R&D spending by Japanese government constitutes only 20% of the total national R&D expenditures. Responding the recommendation of the Science and Technology Council in 1991, Japanese government in the cabinet meeting made a decision to double the R&D budget from the level of 1991 fiscal year as early as possible. I believe the decision must be fulfilled by all means. The doubling of R&D budget is possible, if, for example, the sales tax is raised by 1 %. This policy cannot be put into practice without a support of Japanese citizen. A national consensus among Japanese citizen is necessary.

I hope the new paradigm of science policy and technology management will become a common understanding of our citizen. There will be another effect of doubling the R&D budget, i.e., a continuous investment in R&D as well as public infrastructures will activate our economies. Eventually, it will provide job opportunities for talented youths who are willing to join the science and technology community.

Finally, I would like to emphasize the greatest contribution of Japan to the world should be that Japan take an initiative for the creation of a new society towards the 21st century through science and technology. I hope more young Japanese who find value and pride in science and technology will be engaged in realizing the "Intelligent and Ecological Society" for the coming century.

Thank you very much for your attention.

SCIENCE AND TECHNOLOGY IN INDUSTRIAL DEVELOPMENT IN MALAYSIA

B.C. TAN

Hong Leong Engineering Sdn Bhd
8th Floor, Bangunan Hong Leong 117 Jalan Tun H.S.Lee, 50000 Kuala Lumpur
Malaysia

Abstract. The paper provides examples of the involvement of Science and Technology in Industrial Development in Malaysia. The Malaysian Government believes that the proper application of Science and Technology in Industry is important to fulfill the country's aspiration to achieve advanced status by the year 2020. The Government's support in the drive towards industrialization through Science and Technology is clearly demonstrated.

1. INTRODUCTION

Over the last few years it is clear that the Malaysian economy has shifted from being dependent on agriculture and primary products to one where manufacturing is dominant, and growing steadily. In order for Malaysia to fulfil its aspiration to become an advanced, affluent nation by the year 2020, it needs industrial technology so that it will be able to expedite and consolidate the process of industrialization. Malaysia wishes to understand, assimilate and finally generate its own indigeneous technology.

2. A NATIONAL ACTION PLAN

Although in this paper I shall be concentrating on technology originating from Physics, a country's industrial success requires two other technologies as well ; broadly classified as agricultural technology and medical technology. When I talk about Physics, I shall include engineering and chemistry as well because it is my opinion that they originate from Physics.

For effectiveness, technology must be able to permeate all industrial activities. For technology to play a pivotal role in the industrialization process of a country, it is vital that the development of technology is directed to the needs of the business community. The Government, the Private Sector, the Research Community and Society have important roles to play.

2.1 GOVERNMENT

It must provide leadership and vision, to nurture those areas which, because of their long-term nature or the breadth of their scope, cannot be supported by private enterprise.

It must be pointed out that, in Malaysia, we do not yet have companies like IBM, Siemens, Sony, Hyundai and ICI that can self-support huge R&D expenses. Malaysia is technologically underdeveloped and Government assistance is much needed for industrial R&D.

Because the process of industrial technology development is a long, arduous and expensive journey it is necessary for the Government, in consultation with Industry and experts, to be proactive and to come out with a comprehensive National Plan of Action.

2.2 PRIVATE SECTOR
The private sector has to be fully involved, providing the energy, entrepreneurship and clarity of purpose for the success of technology-based enterprises.

2.3 RESEARCH COMMUNITY
Technological ideas and developments stem from the research community in universities, research institutions and private laboratories. Steps have to be taken to ensure that their products/processes are timely, relevant and commercially viable.

2.4 SOCIETY
It is essential that society recognises the impact that S&T can have on the quality of life. Through education, society can make a value-added judgment and this can be translated, through the political process, into support for proper industrial development.

CABINET COMMITTEE ON SCIENCE & TECHNOLOGY

TERMS OF REFERENCE	MEMBERSHIP
• To examine and evaluate current policies, strategies and programmes on S&T and to ensure that they are in line with national development goals • To formulate, monitor and evaluate policies, strategies, programmes for the short, medium and long term development of S&T in the country	• Y.A.B. Prime Minister (Chairman) • Y.B. Minister of Science, Technology & the Environment • Y.B. Minister of International Trade and Industry • Y.B. Minister of Education • Y.B. Minister of Finance • Y.B. Minister of Human Resources

SECRETARIAT : MINISTRY OF SCIENCE, TECHNOLOGY AND THE ENVIRONMENT

Figure 1

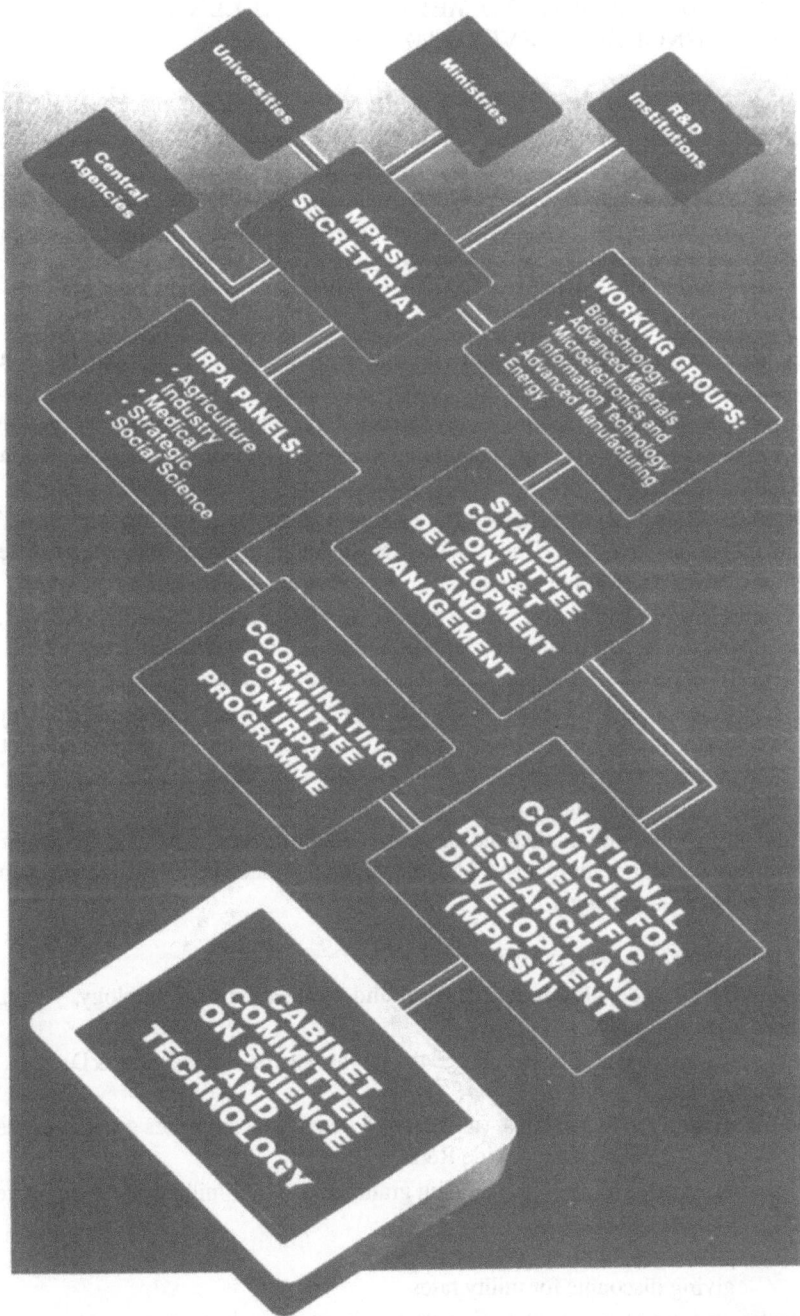

Figure 2

3. STRATEGIC THRUSTS TO MEET THE CHALLENGES FOR TECHNOLOGY DEVELOPMENT

Technology development results more from the pull of the market place than the push of scientific knowledge. Worldwide, the complexity, selectivity and specialization in technology is increasing. Technology has become just about the most profitable business in the developed countries. It is a strategic commodity and there is no natural flow from a higher to a lower level. As a result the Government of Malaysia decided that the process of technology development has to be carefully guided.

3.1 LEADERSHIP
Provide leadership to strengthen the institutional and support infrastructure for Industrial Technology Development.
* A permanent Cabinet Committee on Science and Technology headed by the Prime Minister was set up to ensure committed leadership and strong policy direction. See Figure 1.
* MOSTE was given the responsibility of coordinating R&D programmes in S&T and disbursing funds. This is being done through the Intensification of Research Priority Areas (IRPA) mechanism.
* A National Council for Scientific Research and Development (MPKSN) was set up to ensure effective dialogue, joint planning, consultation and coordination between Government, Industry and Research Communities. See Figure 2.
* The Malaysian S&T Information Centre (MASTIC) was set up to facilitate rapid and effective dissemination of information on R&D within the country, and later outside the country.
* The Malaysian Technology Development Corporation was set up to transfer technology ideas into marketable and profitable projects. This is a private limited company set on a venture capital basis with equity taken up by the Government and private sector companies.

3.2 R&D AND THE PRIVATE SECTOR
Ensure and encourage widespread diffusion and application of technology, leading to enhanced market-driven R&D.
* Incentives are provided to encourage the private sector to invest in R&D, for example,
    - granting pioneer status (no tax for 5 years) to companies specifically set up to conduct industry - wide R&D
    - offering soft loans or matching grants up to RM1 million (RM1 = 40 yen) ; good for SMIs.
    - providing land for R&D facilities
    - giving discounts for utility rates
* Quality and productivity awareness are regulated through standards and programmes in statutory bodies
    - Standards and Industrial Research Institute of Malaysia (SIRIM)
    - National Productivity Centre (NPC)
    - A Centre will be set up for Product Design and Development and a special

THE NEW AND EMERGING TECHNOLOGIES

■ ADVANCED MANUFAC-TURING TECHNOLOGY (AMT) refers to the application of advanced techniques of management and technical methods and methodologies to enhance the quality, speed, and flexibility of the manufacturing environment. AMT draws upon a wide range of technologies, notably information technology and control and software engineering. Examples include CAD-CAM, CNC, industrial robots, flexible manufacturing systems, process planning and control, and expert systems. AMT is applied to varying degrees in some of the larger manufacturing enterprises in Malaysia, particularly those associated with MNCs. Most indigenous manufacturing companies however have yet to derive the benefits of efficiency, productivity, quality and international competitiveness that AMT can offer.

ADVANCED MATERIALS research covers a wide range of materials: plastics, metals and ceramics, and composites of these materials, where recent technological progress has shifted radically the traditional perceptions of the roles that these materials can play. Thus, for example, new structural applications of plastics and ceramics are foreseen; the mechanical and chemical properties of metals are being enhanced through advances in metallurgy and fabrication methods, and whole new class of electro-ceramics are seen as the key to advances in electronics and micro-electronics. Malaysia possesses a resource advantage in plastics and in certain new metals.

BIOTECHNOLOGY has grown in importance as a key technology of the future through recent advances in molecular biology or genetic engineering. Applications are evident in disparate areas, including agriculture, health, food, energy, and biotechnology also has potential applications in industry. This is a field where Malaysia has acquired a competent and moderately successful knowledge base which is well placed to embark upon an intensified thrust as a key technology.

ELECTRONICS is generally recognised as the technology which has had the greatest impact on economy and society within this century. A wide range of specialised areas of activity are encompassed, including the design and manufacture of components, techniques of design and assembly of systems, and a burgeoning range of increasingly sophisticated applications for industry, defence, and the consumer. Micro-electronics and digital technology are two of the key areas in this expanding field. Malaysia has acquired a measure of familiarity with the electronics sector, via the manufacturing activities of MNCs, but has still to acquire a significant level of competence and innovative capabilities in this area.

INFORMATION TECHNOLOGY (IT) in its broadest sense, refers to all technological elements that enable the handling, treatment, and usage of information. The world has experienced a veritable explosion in concepts and facilities to handle information, and IT has aptly been described as the foundation of the second industrial revolution, and has been targeted by virtually all industrialised and industrialising economies as a strategic competitive weapon for attaining efficiency and competitiveness in world trade. IT is generally perceived in terms of the advances in computers and communications. Malaysia, like the rest of the world, is currently experiencing considerable change as a result of the impact of IT, and has demonstrated abilities in certain areas of micro-computer technology and telecommunications, although as a whole it lags behind all NICs in these areas.

*Source: Industrial Technology Development.
A National Plan of Action, Ministry of Science, Technology and the Environment, Malaysia, 1990.*

Figure 3

Technical Committee to propose measures to upgrade the Engineering and Technical Services sectors.

3.3 THE LAUNCHPAD

Build competency in key emerging technologies and commercialise.

* Develop a knowledge base in five selected technology areas to support Malaysian industry (See Figure 3) :
  - advanced materials
  - electronics
  - information technology
  - advanced manufacturing technology
  - biotechnology

These specific areas were selected based on relevance, availability of a natural advantage and constraints of manpower and budget.

DEVELOPMENT ALLOCATION FOR SCIENCE AND TECHNOLOGY (1986 – 1995)

	(M$ million)	
	5 MP Allocation	6 MP Allocation
Direct R&D Programmes:		
Agriculture	203.2	273.8*
Industry	138.1	177.7
Medical	33.1	59.8
Strategic	39.4	78.6
Social Science	—	10.1
Total Direct R&D	413.8	600.0
S&T Infrastructure and Development	126.7	560.3
TOTAL S&T ALLOCATIONS	540.5	1,160.3

NOTE :
  * Allocation does not include cess funds under the Rubber Research Institute of Malaysia (RRIM) and Palm Oil Research Institute of Malaysia (PORIM).

Source : Sixth Malaysia Plan Report, Government Printers 1991

Figure 4

3.4 THE ULTIMATE RESOURCE

Strengthen the institutions and mechanisms for continual development and elevation of technical proficiency of the human resource base.

* A Human Resources Development Fund to finance industry training programmers has been set up.   Manufacturing industries contribute 1% of payroll to the Fund.

* Enhance and institutionalise linkages for industrial training between industry and educational establishments. Courses at institutions of higher learning should include a high degree of exposure to practical training opportunities.

* Strengthen the role of tertiary institutions in advanced technology research and innovation.

LIST OF INSTITUTIONS & UNIVERSITIES FUNDED UNDER THE IRPA MECHANISM

Ministry	Name of Organization
1. Ministry of Housing and Local Government	Housing Research Division
2. Ministry of Primary Industries	Forest Research Institute of Malaysia Mines Research Institute of Malaysia Palm Oil Research Institute of Malaysia Rubber Research Institute of Malaysia Sabah Forestry Department Sarawak Forestry Department Malaysian Cocoa Board
3. Ministry of Agriculture	Fisheries Research Institute Veterinary Research Institute Malaysian Agricultural Research & Development Institute Sabah Agricultural Department Sabah Fisheries Department Sabah Veterinary Department Sarawak Fisheries Department
4. Ministry of Science, Technology and the Environment	Standards and Industrial Research Institute of Malaysia Malaysian Centre for Remote Sensing Nuclear Energy Unit Malaysian Institute for Microelectronics System
5. Ministry of Education	MARA Institute of Technology Universiti Kebangsaan Malaysian Universiti Malaya Universiti Pertanian Malaysian Universiti Sains Malaysian Universiti Teknologi Malaysian Universiti Utara Malaysian
6. Ministry of Health	Institute of Medical Research
7. Ministry of Works	Public Works Research and Training Institute

Figure 5

IRPA FUNDING UNDER SIXTH MALAYSIA PLAN

Institutions	R&D Allocation Recommended By IRPA Panels (M$)					Total	Percentage
	Agricultural Sector	Industrial Sector	Medical Sector	Strategic Sector	Social Science Sector		
1. UTN	2,000,000	14,052,837	2,896,000	5,540,000	-	24,488,837	4.16
2. MIMOS	-	9,000,000	-	-	-	9,000,000	1.53
3. RRIM	17,755,500	10,630,000	-	1,050,000	294,500	29,730,000	5.05
4. PORIM	7,200,000	2,630,000	55,000	-	-	9,885,000	1.68
5. FRIM	7,500,000	13,000,000	-	2,380,000	-	22,880,000	3.89
6. PEGAMA	-	1,100,000	-	170,000	-	1,270,000	0.22
7. LKM	2,965,545	6,509,000	-	505,855	129,000	10,109,400	1.72
8. JPSK	5,000,000	-	-	-	-	5,000,000	0.85
9. JPSB	2,000,000	-	-	-	-	2,000,000	0.34
10. MARDI	127,217,200	10,282,800	-	11,500,000	3,000,000	152,000,000	25.84
11. IPP	27,000,000	-	-	-	-	27,000,000	4.59
12. IPH	3,100,000	-	123,200	-	-	3,223,200	0.55
13. JTSB	1,000,000	-	-	-	-	1,000,000	0.17
14. JISB	1,000,000	-	-	-	-	1,000,000	0.17
15. JHSB	500,000	-	-	-	-	500,000	0.09
16. JISK	1,500,000	-	-	-	-	1,500,000	0.26
17. ITM	-	1,480,000	-	1,000,000	3,629,500	6,109,500	1.04
18. UKM	9,462,000	7,760,000	9,399,822	11,495,040	186,160	38,303,022	6.51
19. UPM	44,869,011	9,301,192	910,000	12,426,120	1,258,547	68,764,870	11.69
20. UM	4,450,000	10,200,000	13,817,250	14,408,000	1,437,500	44,312,750	7.53
21. USM	9,000,000	12,450,000	15,275,000	12,348,120	170,000	49,243,120	8.37
22. UTM	300,000	11,970,000	-	5,080,000	-	17,350,000	2.95
23. SIRIM	-	35,558,000	-	-	-	35,558,000	6.05
24. PRSN	-	-	-	360,000	-	360,000	0.06
25. IMR	-	-	17,353,750	-	-	17,353,750	2.95
26. BPP	-	-	-	335,000	-	335,000	0.06
27. ILP	-	9,900,000	-	-	-	9,900,000	1.68
TOTAL	273,819,256	165,823,829	59,830,022	78,598,135	10,105,207	588,176,449	100.00
PERCENTAGE	46.56	28.19	10.17	13.36	1.72	100.00	

Figure 6

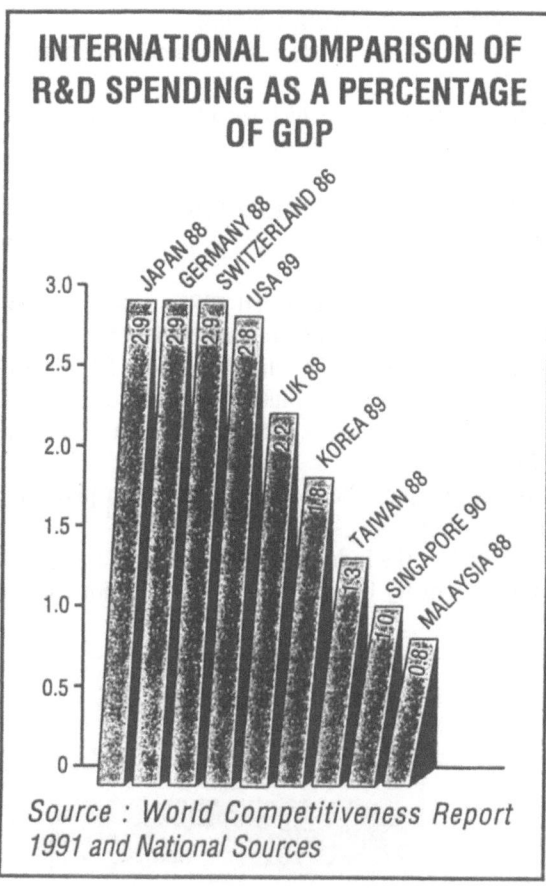

Figure 7

This could be achieved through :

- providing special development budget allocations in areas related to targetted new and emerging technologies
- increasing the number of post-graduate students majoring in the technical disciplines. to 10% of the undergraduate population
- increasing the industry orientation of post-graduate courses related to S&T

* Ensure an effective role of institutions of higher learning in all proposed technology parks and innovation centres. Universities must adopt a more commercial stance in developing technologies.

3.5 SCIENCE AND TECHNOLOGY CULTURE

Elevate S&T awareness and appreciation to provide the most conducive climate possible for invention, innovation and technological advancement.

- in all levels of government
- at all levels in the education system
- through the mass media
- through public programmes such as S&T week

- by strengthening the system for management of intellectual property rights, and enhancing patent advisory and other services
- Malaysia has even included as one of five principles or tenets of nationhood the following :

"To build a progressive society which shall be oriented to modern S&T".

All school children are required to study these tenets.

4. THE INTENSIFICATION OF RESEARCH IN PRIORITY AREAS (IRPA) PROGRAMME

During the Fifth Malaysia Plan (5MP) period 1985 - 1989, the Government introduced the IRPA mechanism with the aim of ensuring selectivity and quality R&D activities in the public sector. For the 6MP (1990 -1994) the allocation is RM600 million. See Figure 4.

Presently, a total of 28 research institutions and universities receive IRPA funds - 6 statutory research institutions, 6 universities and 16 government research institutions. See Figure 5 and Figure 6.

It is interesting to compare the R&D spending as a percentage of GDP in some countries. See Figure 7.

Two sectors, namely, the Industry Sector and the Strategic Sector include some Physics/Engineering /Chemistry based projects. We shall look into these two sectors in greater detail. See Figure 8.

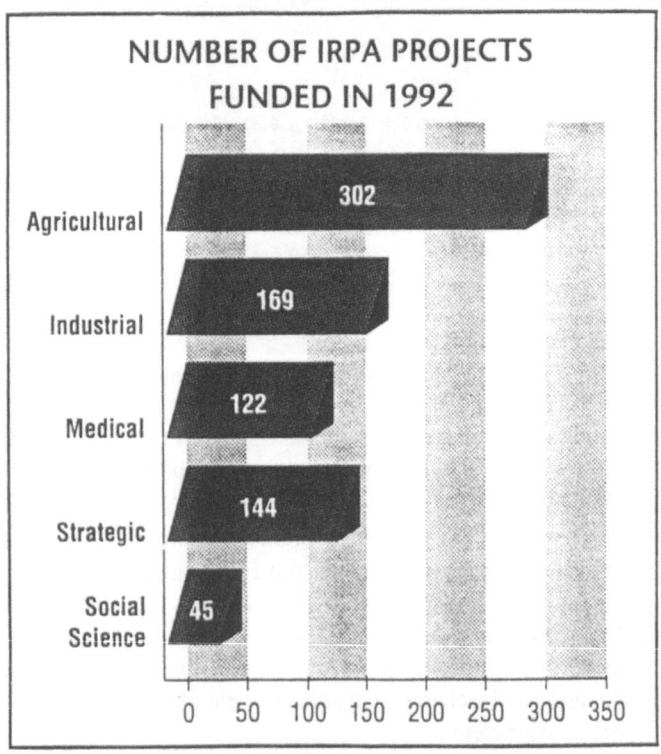

Figure 8

4.1 INDUSTRY SECTOR

Priority areas identified for the 6MP were as follows : -

* Product design and development
  * industrial and consumer products
  * machineries
  * instruments
  * building and structural materials
* Process development
  * process studies
  * production process R&D
  * instrument and control
* Production efficiency
  * improvement in manufacturing techniques
  * improvement in equipment, machinery and plant design
  * production cost reduction
* Quality improvement
  * quality management system
  * instrumental and control
  * standards
  * testing
* Advanced/Enabling Technology Manufacturing Systems
  * automation
  * flexible manufacturing system
  * computer integrated manufacturing

Research conducted covered the following technology areas :

* Materials technology
  - development and usage of metals, ceramics, plastics, polymer and composites for the local industry
* Manufacturing Technology
  - development of value added products from local raw materials, such as rubber, woods, minerals, palm oil, cocoa and food
* Chemical Technology
  - processing, modification and synthesis of rubber, palm oil, cocoa, tin, agricultural waste as well as treatment and product recovery from agricultural and industrial wastes
* Information Technology and Microelectronics
  - development of computer software, telecommunication network, CAD/CAM, expert system, maintenance of equipment, instrumentation and information system
* Industrial Biotechnology
  - microbial metal recovery from industrial and agricultural wastes

INDUSTRY SECTOR PROGRAMMES 6MP
(PHYSICS/PHYSICS RELATED)

Programmes	Allocation (RM)
Metals and Metallurgy	4,110,000
Ceramic Technology	5,380,000
Metrology	800,000
Plastic Technology	1,240,000
Radioactive Tracer Techniques	981,000
Standardization of Radiation Dosage	1,713,000
Narural Rubber Processing by Radiation	1,713,000
Development of Scientific Equipment	931,000
Use of Radiation in the Plastics Industry	1,754,500
Radioactive Sources in Industry	2,218,000
Food Preservation using Radiation	1,000,000
Heterogeneous Computer Network System	5,000,000
Wood-based Research	13,000,000
Rubber-based Research	11,000,000
Strength of Plastics and Other Materials	1,200,000
Molecular Electronics	1,686,000
Thin Film Technology	500,000
Organotin Compounds	1,900,000
Lasers and Electro-optics	1,500,000
Pulse Technology	600,000
Computer Integrated Manufacturing	1,100,000
Polymer-based Industry	1,400,000
Equipment Automation in Industry	1,200,000
Ceramic Materials	600,000
Electronic Materials	800,000
Compressed Gas Technology	500,000
Alternative Energy	700,000
S&T of Materials	1,200,000
Micro-electronics and Telecommunications	600,000
Vibration and Corrosion Studies	1,350,000

STRATEGIC SECTOR PROGRAMMES 6MP
PHYSICS/PHYSICS RELATED

Programmes	Allocation (RM)
Environmental Monitoring	1,500,000
Environmental Management	1,400,000
Solid State Science & Related Technology	1,200,000
Plasma Physics and Laser Plasma Interactions	1,200,000
Fabrication Techniques in Materials Science	150,000
Low-level Radioactivity	600,000
Laser Related Research & Technology	1,200,000
Molecular Electronics	500,000
Magnetic and Semiconductor Materials	600,040
Microwave Equipment	300,000
High Temperature Superconductors	400,000
Renewable Energy Resources	120,000
Environmental Conservation	500,000
Energy Studies	560,000
Radiation Physics	300,000
Atmospheric Studies	750,000
X-ray in Industry	1,000,000
Efficcienct Use of Energy in Buildings	800,000
Radioactivity in the Environment	2,080,000
Semiconductor Studies Using Radiation	540,000

LIST OF RESEARCH RESULTS FOR COMMERCIALISATION

NAME OF PROJECT	INSTITUTIONS
1.Radiation vulcanised natural rubber latex	UTN
2.Carbon black masterbatch from field latex	RRI
3.Special heat-resistant Newcastle disease (ND) vaccine	UPM
4.Utilisation of modified peat soil in local industries	USM
5. Rapid diagnosis of viral hepatitis	USM
6.Rapid diagnosis of typhoid fever	USM
7.Coordination chemistry and separation of transitional metals from by-products from tin mining	USM
8.Tracer technique in industry	UTN
9.Radiation calibration	UTN
10.Radiation indicator	UTN
11.Surface coating using ultra-violet and electron beam	UTN
12.Radiation sterilisation of medical products	UTN
13.Food irradiation	UTN
14.Computerisation of healthcare, medicine, transformation system & poison related information	USM
15.Temporal bone holder	UKM
16.Noise assessment software	UKM
17.Thermoplastic natural rubber	UKM
18.Compact and simple treatment module for electroplating wastewater	UKM
19.Expertise on aquaculture industry	USM
20.Exploitation of shrimp skin, crab and king-crab shell into chitin and chitosan	UKM
21.Measurement of multi-component flow using electrical charge tomography (ECT)	UPM
22.Recovery of carotenoid	PORIM
23.Blackboard and the process of manufacture from oil palm trunk	PORIM
24.Stabilisation of palm oil and palm oil fractions	PORIM
25.Esterification of carboxylic acid/glyceride mixtures	PORIM

4.2 STRATEGIC SECTOR

Indicative areas for the sector in the 6 MP were as follows :

*   National Resources Management
    *   ○   Remote sensing
    *   ○   Marine Sciences / Oceanography
    *   ○   Earth Sciences
    *   ○   Energy
    *   ○   Hydrology
*   Environmental Management
    *   ○   Environmental studies including resource recovery
    *   ○   Climatology
    *   ○   Conservation of Genetic Resources
    *   ○   Environment Risk Assessment
*   Foundation Sciences and Technology
    *   ○   Basic Sciences
        *   -   Physics, Chemistry, Mathematics, Biology, Genetics
    *   ○   Emergent Technologies
        *   -   Biotechnology, Materials, AI, Lasers and Electro-optics, Information Technology
    *   ○   Techno-Economic and Social Science Research

4.3 FOLLOW-UP REPORT

A significant percentage of research - mainly in universities - is developing in connection with environmental issues. Research oriented towards "new and emerging technologies" accounts for 11% of total IRPA allocations, and within this category two-thirds is allocated to biotechnology related areas.   MPKSN's objective of steering Malaysia's research towards industrial application is evident but support of new scientific and technological frontiers, apart from biotechnology, is thin.

The ability of IRPA funded institutions to absorb new areas of funded research is constrained for information and communication technologies and in new and emerging fields.

One difficulty that was identified as being a factor in low absorptive capability was that levels of research capability in engineering and technology were low across all sectors and all types of institutions.

There are signs that the authorities are determined to pursue R&D in the newly emergent technologies.   It is clear that there are insufficient number of scientists with the required ability to utilise funds set aside for the stated purpose.   Since this bottom-up approach did not succeed, there are now clear signs that a top-down approach may be taken.   This means that a decision may be made t o import this capability and facilities will be set up and funds set aside to ensure the implementation of this strategy.   See Figure 12.

86

ADVANCED MATERIALS RESEARCH CENTRE (AMREC)

AMREC will act as the centre of excellence for scientific and technological research on advanced materials, and as a national focal point for coordination of international activities on materials research and development.

The major functions of AMREC are:

o   to undertake scientific research on advanced materials with the objectives of understanding the properties, optimising required properties, identifying new properties and developing new materials;

o   to undertake technological research on advanced materials with the objective of developing processes and techniques related to the production, fabrication and new applications of advanced material;

o   to undertake joint R&D programmes with local and foreign institutions; and

o   to establish and up-date information on advanced materials.

The major thrust areas of AMREC will be:

o   materials characterisation and evaluation;
o   materials design;
o   materials processing; and
o   materials system and product development

To undertake the above R&D activities, it is proposed that AMREC will have the following specialised laboratories:

o   Materials Science Research Laboratory;
o   Materials Engineering Research Laboratory;
o   High Temperature and Pressure Laboratory; and
o   Pilot Plant and Processing Laboratory.

Figure 12

4.4 A PROPOSAL FOR AN OPTOELECTRONICS MATERIALS AND DEVICES LABORATORY

4.4.1 INTRODUCTION

As some of you may know, I was involved for some years in laser research. We built several types of lasers and used them in some applications. I believe that the work should be continued under the area of Advanced Materials for the following reasons.

Of all the emerging technologies, optoelectronics is one of the fastest expanding areas both in terms of R&D output and revenue generated from sales of optoelectronics related products. Optoelectronics concerns the production of photons and their manipulation with or without associated involvement of electronics, in a media which is usually in a solid phase. Through exploitation of some very fundamental properties of light, quantum leaps in existing technologies are possible with potential new technologies in the horizon.

Among the important by-products of optoelectronics research in the market are the ubiquitous compact discs, laser printers, bar-code scanners, lasers and diode lasers, LEDs and luminescent displays and optical fibres. Some newer optoelectronic technological products being pursued at the moment are optical computers, higher power tunable lasers, optical memories, optical sensors and optical fibre amplifiers. While many marketable products have arisen out of optoelectronics, the fundamental scientific work has by no means ended and much remains to be done in many areas especially in areas related to interaction of light with a chosen media. Since photons are anticipated in the 21st century to be what electrons have been in the 20th century, optoelectronics related materials will represent in the true sense of the word "Advanced Materials".

There is a need for the development of a basic facility for the fabrication and production of optoelectronics materials and devices, in complement to the existing programme of lasers and optoelectronics research currently conducted at the University of Malaya. In order to be able to cut short the lead time in the development of this field, it is suggested that an interested party, already in this research area, get involved in this venture through Prof. Kum Sang Low of the University of Malaya.

4.4.2 OBJECTIVES

The objectives of the Optoelectronics Materials & Devices Laboratory are :
- to develop specific areas of research in optoelectronics
- to develop and keep pace with existing and emerging optoelectronic related technologies
- to provide technological assistance to the small but growing optoelectronics industry in Malaysia
- to develop products which could eventually generate revenue to the country
- to provide highly skilled training for future (B.Sc., M.Sc., Ph.D etc.) for the optoelectronics industry

5. MANPOWER DEVELOPMENT

There are eight local universities in Malaysia, namely :

Universiti Sains Malaysia
Universiti Pertanian Malaysia
Universiti Teknologi Malaysia

Universiti Kebangsaan Malaysia
Universiti Malaya
Universiti Utara Malaysia
Universiti Sarawak Malaysia
Universiti Islam Antarabangsa

Between them and also counting those returning from abroad the number of Physics graduates produced per year are approximately :

B.Sc   -   180
M.Sc.   -   12
Ph.D.   -   4

R&D MANPOWER IN THE PUBLIC SECTOR

■ The findings of a R&D manpower survey undertaken by the Ministry of Science, Technology and the Environment in 1990 revealed that out of a total of 22013 research personnel and supporting staff in public sector R&D institutions and universities, 6504 or 29.5% were research scientists and the rest were supporting staff. Of this number of research scientists, some 1743 (27%) are Ph. D holders, another 2665 (41%) possess Masters qualifications and the rest (2096 personnel or 32%) are first degree holders.

An earlier survey undertaken by the Ministry in 1989 revealed that the public sector accounted for almost 87% of the total number of R&D personnel including supporting staff in the country, a substantial portion of whom were engaged in agriculture research. A large proportion of the R&D personnel in the public sector are specialised in basic and agriculture sciences compared with that in the applied or engineering disciplines. This uneven representation in the public sector coupled with the small number of R&D personnel in the private sector has contributed to the consequent low impact of R&D on industry.

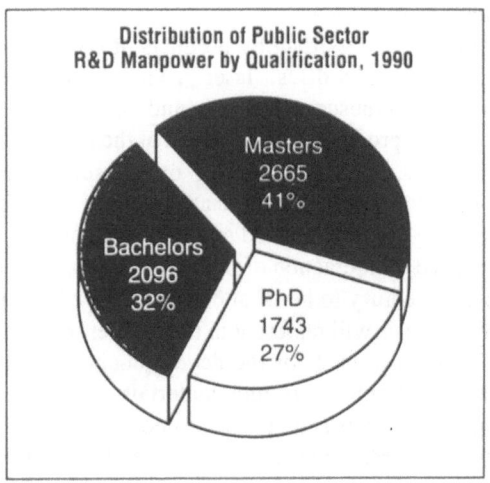

Distribution of Public Sector R&D Manpower by Qualification, 1990

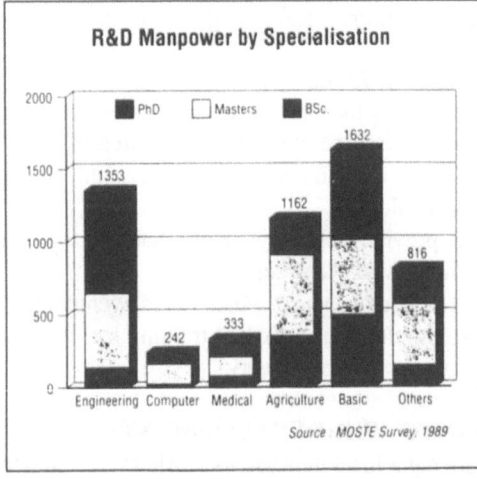

R&D Manpower by Specialisation

Source : MOSTE Survey, 1989

Figure 9

The Government has lately encouraged the setting up of twinning arrangements between local private institutions and foreign universities. This arrangement will assist in producing more first degree graduates. However, postgraduates will continue to be trained in the local universities and abroad but the number will remain low.

Figure 9 shows the R&D Manpower (1990) in the public sector. Out of the total of 1743 Ph.D holders, about 200 are physicists and out of the 2665 M.Sc. holders above 250 are physicists. A large proportion of the total R&D personnel are engaged in basic and agricultural sciences compared with that in the applied and engineering disciplines. This fact contributed to the consequent low impact of R&D on industry.

The proportion of workforce employed in R&D activities in Malaysia is very small. There are currently about 7000 full-time research scientists and engineers in the country. This is about 10 research scientists per 1000 workforce. This much lower than that found in the developed countries and in certain NICs. See Figure 10.

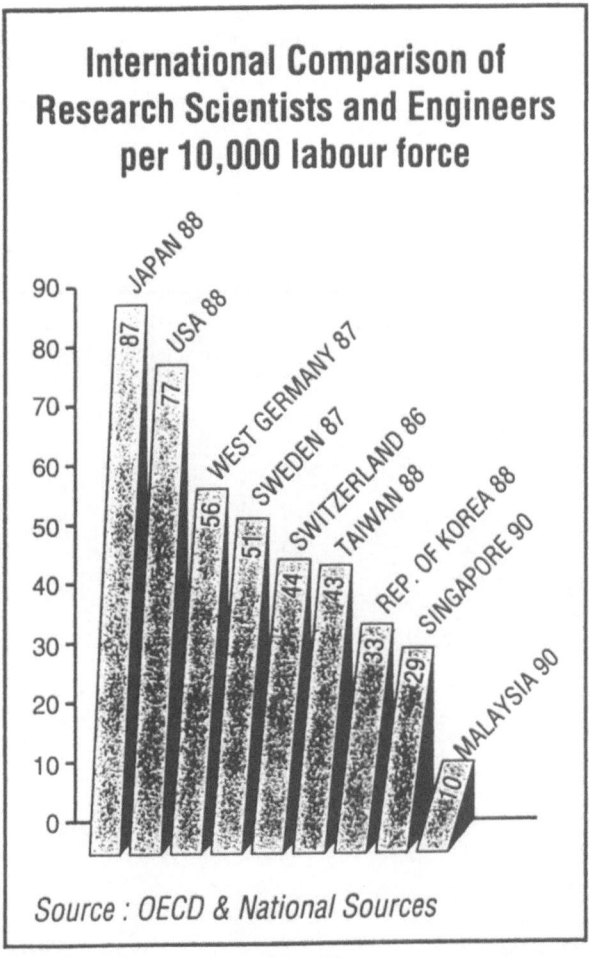

Figure 10

THE NEWLY ESTABLISHED CONSULTANCY UNITS AT LOCAL UNIVERSITIES

■ UNIVERSITI TEKNOLOGI MALAYSIA

RESEARCH AND CONSULTANCY UNIT

Dean	:	Ir Dr Zainal Aripin Zakariah
Deputy Dean (Research)	:	Assoc Prof Dr Mohd Nor Musa
Deputy Dean (Consultancy)	:	Encik Elias Salleh

Major Research Areas for Collaboration
* Electronics
* Electrical
* Materials
* Engineering
* Environmental Impact Assessment
* Vibration

■ UNIVERSITI KEBANGSAAN MALAYSIA

RESEARCH AND CONSULTANCY BUREAU

Director	:	Prof Dato' Dr Nik A Rashid Ismail
Deputies	:	Dr Sharifah Mastura Syed Abdullah
		Dr Ahmad Badri Mohamad

Major Research Areas for Collaboration
* Consultancy and Testing
* Problem Solving
* Training Courses

■ UNIVERSITI PERTANIAN MALAYSIA

CONSULTANCY INSTITUTE

Director	:	Encik Khalid Rafa

Major Research Areas for Collaboration
* Agriculture
* Resource-based research
* Engineering
* Environmental Projects

■ UNIVERSITI SAINS MALAYSIA

INNOVATION AND CONSULTANCY CENTRE

Director	:	Prof Dr R Ratnalingam
Consultant	:	Dato' Chet Singh

Major Research Areas for Collaboration
* Pharmaceuticals
* Polymer
* Food
* Machinery and Engineering
* Automation
* Environmental Control
* Chemical
* Aquaculture
* Medical Diagnostic Kits

■ UNIVERSITI MALAYA

CONSULTANCY UNIT

Director	:	Dr Amiruddin Adnan

Major Research Areas for Collaboration
* Environmental control/waste treatment/agro-waste
* Engineering - Automation, robotics, computer research control
* Botanical/geological studies
* Laser Technology
* Petrochemical
* Advanced Materials
* Corrosion
* Separation Processes
* Pharmaceuticals

Figure 11

6. UNIVERSITY - INDUSTRY LINKAGE

A recent survey on prevailing university - industry linkage in Malaysia concluded that

○   universities were generally unaware of R&D needs of industry

○   the industry similarly was unaware of the expertise, facilities, findings and studies being conducted by universities ; and

○   industry, particularly SMIs, was unable to identify their R&D and technology needs.

Based on the findings, MPKSN initiated measures requiring public research institutions and universities to take appropriate steps to ensure greater interaction.   See Figure 11.

7. CONCLUSION

In the early 70s and 80s, mainly US, European and Japanese electronic companies moved out their more labour intensive operations to the Far East to take advantage of lower labour costs and tax incentives available.   The local companies were involved in the processing of silicon wafers, wire bonding and encapsulation of IC chips into packages and in quality testing.   Malaysia has a large share of this business because the labour was good and cheap and the country's infrastructure was decent.

In the last few years, a number of these electronic companies are moving their lower IC wafer fabrication plants to Malaysia.   These are mainly for discrete components and LSI devices.   To support this development will require solid state physicists, laser and optoelectronics physicists, materials physicists, etc.   To a certain extent, the local universities are responding to the challenge to provide the manpower and the basic physics training.

Apart from electronics, the other technological areas that are developing in Malaysia and in the ASEAN region that will need physicists would include telecommunications, automotive industry, medical physics, information technology and materials.

There is definitely a general lack of indigeneous high technology R&D in Malaysia. Our development is dictated by world economic forces.   Even though, at one time or other, we were large commodity producers of tin, rubber, palm oil, tropical hardwoods and petroleum, no high-technology development linked to these commodities ever resulted.

Today we are in danger of losing our semiconductor industry to countries like China and Vietnam because labour costs are rising in Malaysia.   We need a concerted effort to maintain our GDP growth.   We must attract and phase in the higher end production processes of IC fabrication.   We have to automate our factories and aim for higher productivity.   Industry needs more R&D people and certainly more physicists.

BIBLIOGRAPHY

1. Industrial Technology Development : A National Plan of Action.   Report of the Council for theCoordination and Transfer of Industrial Technology.   MOSTE, Malaysia, Feb. 1990.
2. National Council for Scientific Research and Development, Malaysia. Annual Reports 1990 -1991 and 1992.
3. Direktori Maklumat IRPA - MOSTE, Malaysia, 1991.

PHYSICS IN INDUSTRY :
THE HISTORICAL CONTEXT OF CURRENT EVENTS

Morrel H. Cohen
Corporate Research Laboratories
Exxon Research & Engineering Company
Annandale, New Jersey 08801 USA

ABSTRACT. Industrial research in physics and in other sciences has changed its character over the last two decades. The significance of these changes can best be understood in the context of the evolution of science, advanced technology, and national economies over the last three and one-half centuries. For the first time, science has come into dynamic "*equilibrium*" with the economy as a whole, with consequent uncertainties in the optimal management of science, both in society at large and in industry. In my talk I shall describe this historical context, the current state of industrial physics, opportunities for research of benefit to society, and how best to organize research in physics, given current circumstances, to realize those benefits.

I . INTRODUCTORY REMARKS

The character and style of industrial research in physics and in other sciences has changed significantly over the last two decades.

The changes and the constraints they have introduced can best be understood in the context of the co-evolution of science, of advanced technology, and of national economies over the last three and one-half centuries.

For the first time, science has come into "*dynamic equilibrium*" with the economy as a whole. There are consequent uncertainties as to the optimal management of science, both in society in general and in industry in particular.

I shall introduce the subject of change in the circumstances of physicists via a brief personal history. I shall then interpret these personal observations within a framework provided by Derek Price's quantitative studies of the growth of science in relation to the growth of "*high*" technology and of the economy as a whole since the mid-seventeenth century. Next, I shall argue that Price's findings provide a natural framework for understanding the recent changes in the character and style of physics in industry and the slowing of its growth. In doing so, I shall focus on industry in the U.S. because that is where my experience lies. Nevertheless, there are parallels with other countries, and the ultimate driving force for the changes I describe is, I believe, universal. In stark contrast to these changes, which seem strongly negative to those experiencing them most directly, there seems to be continuing growth in opportunities for research in physics appropriate to industry and promising of future benefit to society, and I shall give a few examples. How, then, can we best realize that promise in the face of constraints on growth? I shall conclude by discussing an interesting model which may well be in the early stages of emergence in the U.S. It is a model for the organization of research in physics, emphasizing the industrial role.

II. SOME PERSONAL HISTORY

To dramatize how much career prospects for fresh Ph.D. physicists have changed for the worse over the last four decades, I now recall my own experiences. I entered graduate study in physics at the University of California in Berkeley in the Fall of 1948, becoming one of literally hundreds of graduate students in the Department. Such popularity of physics, replicated in various large research departments around the country, was unprecedented. Its causes were various. Returning veterans of the Second World War who had received some technical training or experience during the war were quite naturally attracted by physics. The contributions of physics and physicists to the war effort were widely and dramatically publicized. There was a related, if vague, sense that jobs were available, that a career in physics was a practical possibility. Indeed, while I was a graduate student, established industrial research laboratories like the Bell Telephone Laboratories, the Westinghouse Electric Research Laboratory, and the General Electric Research Laboratory were expanding their staffs in physics, and new laboratories strongly emphasizing physics, like the IBM Research Laboratories, were starting up.

I had planned to complete my Ph.D. by Fall 1952. In the Spring of that year, Charles Kittel, my Ph.D. thesis advisor, called me into his office to discuss possible faculty positions for me. He quickly ran down a list of possible university physics departments and selected the University of Chicago as most suitable. Astonishingly, I had an offer from Chicago three weeks later! Within a very brief period, Kittel had similarly placed five of his students in appropriate positions.

Once at Chicago, I was able to carry out my research with funding provided by the Institute for the Study of Metals, an autonomous research Institute within the University now named the James Franck Institute. The funds were supplied in large measure by the industrial sponsors of the Institute, medium sized and large companies, which received in return quarterly meetings of their scientific staff with the Institute Faculty and Quarterly Progress Reports containing preprints of all Institute publications. I was promoted regularly and received a tenured position at Chicago after only five years and five publications. I readily found positions for my Ph.D. students upon their graduation in the fifties and early sixties. There were many opportunities for consulting at the major industrial research laboratories as well.

It was not apparent at the time, but I was living through a positive fluctuation highly favorable for physics. That two-decade period from the end of the Second World War to the early to mid-sixties was an anomaly, different from the thirties, a negative fluctuation, different from the war period during which the normal activities of physicists were largely suspended, and starkly and fundamentally different from the present.

In fact, that fluctuation masked an underlying downward trend in the growth of physics and encouraged the continuation, almost to this day, of expectations which were, at best, overly hopeful. This downward trend first became apparent to me, personally, in the dying out of our Industrial Sponsors Program as small companies decided that from it they derived insufficient benefit and large companies decided that they preferred to spend the funds within their own laboratories. Next, even though there were still ample government funds supporting my research and the training of my graduate students and post-docs, it became progressively harder to find positions for them. Finally, by the end of the sixties, competition for government research funds had started becoming progressively more intense.

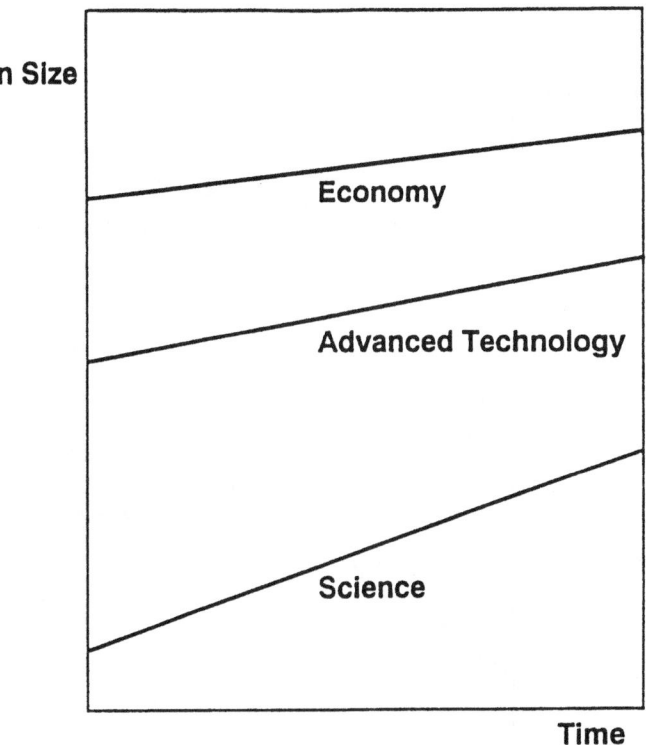

FIGURE 1   A sketch of the comparative growth rates of science, advanced technology, and the economy as a whole [5] .

These negative trends in funding and jobs have continued into the present. In order to place them in a larger historical context, I have returned to what Derek Price, Avalon Professor of History of Science at Yale University before his death, told me in my kitchen in the mid-fifties, long before I could appreciate its significance.

III.DEREK PRICE ON THE GROWTH OF SCIENCE

Professor Price had written extensively on quantitative measures of the size of science and of its growth (see e.g., refs. [1-4] and [6]). His private, informal summary for me of his conclusions as of the mid-fifties is as follows [5] . It became possible to start measuring growth in the size of science from the mid-seventeenth century with the founding of the Royal Society of London. Whatever measure one uses of the growth rate of science -- number of members of scientific societies, number of journals, number of authors, number of papers, shelf space in libraries, etc. -- one obtains the same result: Science grows exponentially with a doubling time of 10-15 years, say about 12 years. One finds similar exponential growth for what one might call the high technology of the time, but now with a doubling time of roughly 24 years. Exponential growth of the world economy has been known for some time; its doubling time is roughly 36 years. Figure 1 shows an illustrative sketch of price's statements on comparative growth.

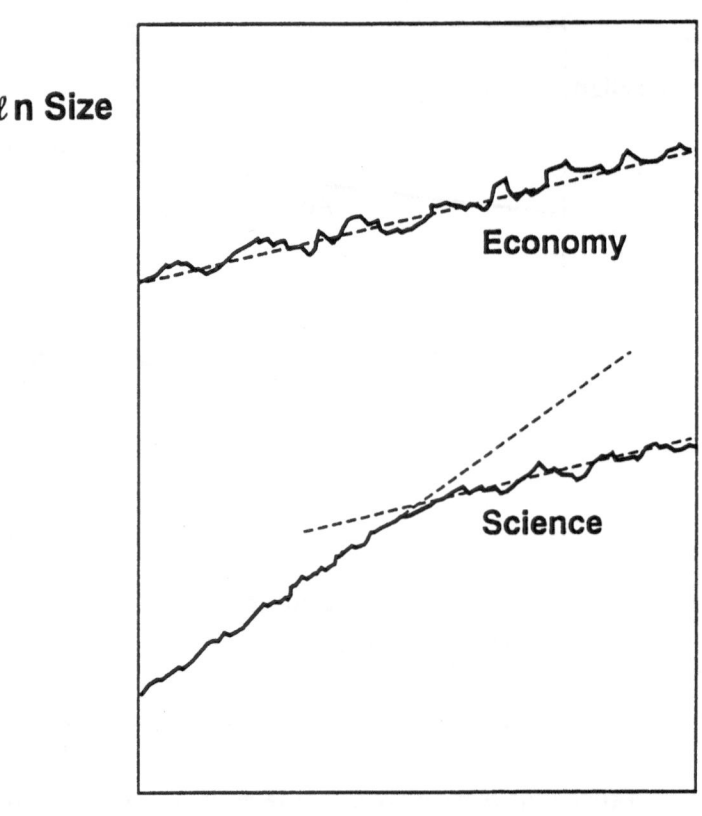

ℓn Size

Economy

Science

Time

FIGURE 2   Dynamic equilibration of science with the economy, a sketch.

These conclusions apply to all sciences, taken collectively. They apply also to individual sciences as well, as shown for example, by the cumulative total of papers abstracted in Physics Abstracts published between 1900 and 1950 [1] . Moreover, while they apply worldwide, they also apply albeit with greater statistical uncertainty to individual countries [3] .

Recalling this conversation with Price a decade later, the explanation for my personal observation of the various manifestations of a decreasing growth rate in physics in the U.S. in the sixties became obvious in retrospect. *It was simply macroeconomics laying its heavy hand on science.* Science could not continue forever to grow twice as fast as advanced technology and three times as fast as the whole economy. At some point it had to come into dynamic equilibrium with advanced technology and with the economy itself.

Once science has grown to the point where its cost becomes comparable to those of other investments which individual firms or entire countries must consider, science must enter the normal processes of competition for funds and be subjected to criteria for prioritization and selection extraneous to science itself. That point presumably occurs first for those countries most active in science in relation to their GNP, those countries

Sources of Industrial R&D Funding

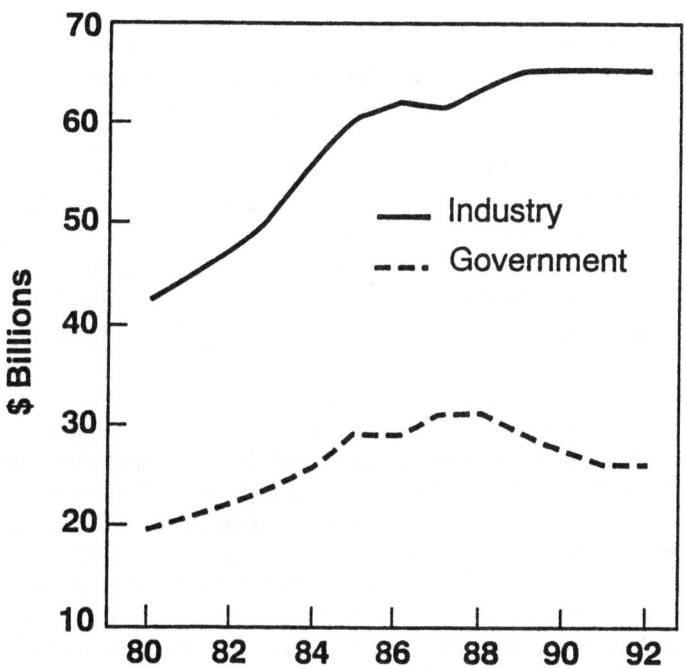

FIGURE 3   Sources of industrial R&D funding in constant (1987) dollars [7] .

near the upper limit of the distribution of scientific activity vs. economic size constructed by Price in ref. [6] for the year 1967, a critical one. It has clearly happened over the intervening quarter of a century for the U.S. and other countries near Price's upper limit (Vid. the figure in ref. [6]).

This is a situation unprecedented in history. We are in the midst of a transition from untrammeled exponential growth of science, doubling every 10-15 years, to growth at roughly the growth rate of the economy as a whole. We have no previous experience to build upon, and no traditions to draw upon. We have entered a period of experimentation, consciously or unconsciously, and we can continue to expect only fluctuations about a growth trend steadily approaching parallelism with that of the global economy, Figure 2. I can hold out no hope for a reversion to the good old days. We must learn to function effectively in what is a new world.

Having discussed in the previous section the manifestations of this slowing of growth, which I perceived in academia from the mid-sixties on, I now describe in somewhat more detail the corresponding manifestations which I perceived more recently in industry.

IV. THE IMPACT OF THE GROWTH OF SCIENCE ON PHYSICS IN INDUSTRY

As stated in the previous section, science has grown to the point where its cost has become comparable to those of other investments which individual firms or entire countries must consider. It has, therefore, entered the normal processes of competition for funds and has been subjected to criteria for prioritization and selection extraneous to science itself. This is true for advanced technology as well. Science and technology have not been effective in that competition, clear evidence for which is shown in Figure 3, which displays National Science Foundation data on sources of industrial R&D funding for the U.S. for 1980 through 1992 [7] . The industry contribution has flattened, while the government contribution has actually decreased (in 1987 dollars). At present, spending on R & D averages 3.7% of total sales [8] , a percentage which is decreasing because R & D support is flat, while sales are slowly increasing in real terms.

Not only is R & D spending leveling off, but funding is shifting away from longer-term investment in basic research towards shorter -term applied research and development. This is manifest in the changes which have occurred in industrial research labs with significant past efforts in basic research in physics. Downsizing, focusing on shorter-term projects, coupling the activities of the laboratory more closely to the *current* activities of the business, even transformation of research laboratories into technical service centers one or more of these has occurred at laboratories of many corporations, among them AT&T, Bellcore, IBM, GE, Westinghouse, Exxon, B.P., Union Carbide, Schlumberger, Monsanto, and others. Such trauma has been the norm for industrial research laboratories in the U.S. Their fortunes wax and wane and sometimes wax again. What is different now is the frequency and stability of the negative changes and the paucity of interesting new laboratories.

There has been much written about these changes in newspapers, magazines, professional journals, even in books, but I have seen no reference to the direct consequences of Price's findings on comparative growth. Explanations abound which focus on details of the current situation and miss the inevitability of an event rooted in the centuries of untrammeled exponential growth of science. Were science still a small endeavor, these oft cited causes would have little effect on its growth.

Nevertheless, some of these causes are significant and worth dwelling on briefly. There is now a trend in U.S. technology companies increasingly to hire non-technical top executives to run their businesses. At present only 18% of American CEO's have technical training vs. 36% for Japanese corporations [8] . In a survey of 300 large companies worldwide, American companies considered 21% of their investment projects long term, i.e., no return expected for the first five years, whereas for Japanese companies the figure was 47% and for European companies the figure was 61% [9] .

To realize the benefits of basic research in physics, steady, long-term investment is needed. There are many well-known examples which illustrate the long time scale required. To cite a familiar one, it was 20 years between Felix Bloch's 1928 Ph.D. thesis describing energy bands in crystals and the first observation of transistor action in 1947 [10] . It was another 20 years before the full impact of that discovery on society became manifest. While it is difficult to predict the return on investment in any particular scientific project, retrospective study has suggested that the long-term rate of return to the U.S. as a nation on its entire national investment in scientific research is 28% per year

[11] , substantially greater than the typical rate of return obtained by American companies from their short-term investments.

There have been many interpretations of this turning away from a successful and valuable long-term orientation by U.S. industry. Among them are the pressures of international competition; the trend towards consolidation; fragmented and transient ownership of companies; the nontechnical top management; the short-term orientation of the companies, financial markets and institutional investors; the relatively high cost of capital in the past; etc. However, the central point is that American companies are highly successful when measured by the criterion of greatest importance in the U.S. financial structure, short-term profitability [9] . The irony is that there is little direct connection between higher short-term profits and long-term capital gains, which have been higher for Japanese shareholders than for American for two decades and higher for German shareholders for one decade [9] . By demanding high near-term profits, the American system is counterproductive, sapping growth by lowering investment to rates far below Japan and Germany [9] . Among investments in such a climate, those in longer-term research are the most vulnerable. Nevertheless, were scientific research still a small and less costly endeavor, it would not receive its present negative scrutiny. Its earlier, rapid exponential growth would persist. We have the misfortune of living in interesting times.

The immediate cause of the slowing of the growth of science in general and physics in industry in particular in the U.S. may, as just implied, be the corporate investment policy prevalent in the U.S., but such slowing is occurring in Japan, Germany, and other countries as well, with different investment policies and quite different immediate causes. The ultimate cause is the same in all such cases, the reaching of some characteristic size. In the initial exponential growth of science, its growth rate depended primarily on its own size. As that characteristic size is reached in each economic unit, the cost of science becomes significant enough to limit its growth rate to that of the economy, the particular mechanisms or immediate causes of limitation differing among nations. That this limitation is occurring now is ironic. The opportunities for the impact of science, in general, and physics, in particular, on the well being of society have never seemed greater, as discussed in the next section.

V. OPPORTUNITIES

In an oft quoted phrase, Vannevar Bush spoke of the "*endless frontier*" of science. "*Endless*" is perhaps an understandable exaggeration, but there is no question that the frontier of physics is now enormous and rapidly expanding. Much of what lies near that frontier is of great potential interest to industry. One can organize those possibilities into trends or themes, several of which I discuss here, in no special order.

Physics is growing more diverse, with proliferating specialties and sub-specialties, but there is unity in that diversity. There is a merging of high-energy physics, or particles and fields, with cosmology. Some of the central ideas of both first emerged in condensed matter physics or statistical mechanics. On the other hand, powerful theoretical techniques and mathematics new to physics first utilized in field theory have found fruitful application in condensed matter physics, statistical mechanics, and newly resurgent areas of classical physics. The use of new experimental techniques and powerful instrumentation quickly diffuses through all of physics. A well-trained, well-informed, and alert industrial physicist can import and build upon relevant developments throughout all of physics.

Computers have continued to increase in power per unit cost by an order of magnitude every five years. Sophistication in the use of computers has kept pace with the increase of power. There has been a concomitant development of both theoretical and mathematical tools so that computations or simulations can be carried out which were beyond dreaming when I started my career. To cite one of many possible examples, the density functional theory of Hohenberg and Kohn, and especially its formulation by Kohn and Sham, has revolutionized the computation of the electronic structure, atomic structure, and properties of crystals, surfaces, clusters, etc. Although these methods were introduced thirty years ago [12] , it is only now that my dream of forty years ago is thereby becoming feasible, the design of materials from first principles, clearly of interest to industry.

One of the recent trends most surprising to a physicist of my generation is the resurgence of classical physics. Fluid dynamics, pattern formation, granular flows, dynamical chaos, self-assembly, nonlinear dynamics, and many other classical fields have become so active that in some physics departments h is set equal to zero more often than it is set equal to unity.

In part, because of this growing interest in classical systems and, in part, because of a growing command of theory and of new experimental tools of great power, precision and subtlety, physicists have had the courage to study systems of greater and greater complexity, systems either of structural complexity, regular or random, or of dynamical complexity, or of both. The materials on which my industry is based are incredibly complex fluids, and my laboratory has played a central role in founding and fostering a branch of physics of that name, "*complex fluids*". Those materials form, in turn, part of a larger group of complex materials which my Laboratory has also actively studied, "*soft condensed matter*", the importance of which has been recognized by the recent award of the Nobel Prize to P.-G. DeGennes..

Throughout much of my career, the major focus in physics departments was usually two-fold, on the sub-nuclear world of high-energy physics and the macroscopic world of condensed matter physics. In some departments there was also an effort in atomic and molecular physics, but typically on a smaller scale. Now there is a growing fascination with the excitement of the small, the territory lying between condensed matter and atomic/molecular physics. Mesoscopic physics, cluster physics, the physics of single atoms, and molecular electronics all lie in this territory. We now have the ability to manipulate individual atoms. We have a broad range of visualization tools that enable us to "*see*" at the atomic scale. We can even "see" single isolated atoms. Single-electron devices have shown transistor action. Interesting new physical phenomena are emerging all over this new territory. There is reason to expect, or at least to hope, that large-scale integrated devices can ultimately consist of densely-packed atomic scale units. Even the old barrier to the dense-packing of devices, heat dissipation, may fall with new ideas on quantum computation. Until now, most devices have depended on the manipulation of charges or of photons coupled to charges, with magnetism playing a secondary role, except in certain memories and tape devices. Now, however, transistor action based on spin manipulation has been demonstrated in a thin film device. The fabrication technology developed by industry for large-scale integration has made possible the investigation of mesoscopic and smaller scale physics. It has also branched out in novel directions, the fabrication of micromachinery, for example.

Complex materials and structures often have complex constituents and properties Effective research often requires collaboration with inorganic chemists, organic chemists, solid-state chemists, polymer chemists, synthetic chemists, materials scientists, fluid dynamicists, chemical engineers, mechanical engineers, electrical engineers, mathematicians, and/or computer scientists. Such interdisciplinary collaborations are often easier to bring about in an industrial setting, where organization units can be constructed along functional, rather than disciplinary, lines. However, few institutions, be they academic, governmental, or industrial, can command all of the skills needed effectively to address the many present opportunities. Inter-institutional collaborations are growing, and industrial organizations are learning how to deal with the attendant intellectual property issues. I shall discuss such organizational questions in the next section.

Biomimetics is an interdisciplinary area receiving increasing attention, and I shall comment on two examples here. The first is learning how to fabricate structural materials of great strength or high strength to weight ratio through the analysis of such biological materials as spider web, bone, shell linings, chitin, etc. The second concerns learning from the structure and organization of the human brain how to reach further in the design of computers. The emergence of the so-called neural networks from a fusion of spin-glass physics with simple ideas about how real neurons are connected is a familiar example.

There is much more to be learned, however. We now have parallel computers which run the gamut from few complex processors to many simpler processors. We also have special purpose computers, hard-wired to do specific tasks efficiently. The brain has all of this and more: many more neurons than transistors in our fastest computers (10^3 times); massive parallelism with connectivity as much as two orders of magnitude higher than anything we have achieved; many highly specialized sub-computers linked together; complex individual processors; plasticity; many signaling agents, both ions and molecules; both point-to-point, regional, and global communication; multiple special purpose memories; enormous speed for certain very complex tasks, despite the slowness of the individual neurons and of interneuronal communication. There is great opportunity here for physicists, both in the biological physics and in the territory of the small discussed above.

Given the present constraints on the growth of physics as a whole and the constraints on physics in industry, how then do we address these opportunities? How do we initiate new enterprises when there is difficulty carrying on enterprises of current importance?

VI. A MODEL FOR RESEARCH IN PHYSICS

I shall now outline briefly a model for research in physics designed to improve the effectiveness of research in industry in contributing to business success while maintaining the health of the overall research enterprise in the new slow-growth environment. None of its essential elements are new; all are present in one form or another in various countries. What is different here is emphasis and integration.

First, industrial organization must make a clear distinction between precompetitive or generic research and proprietary research. The former may be basic or it may be something conventionally described as applied, but it is usually too general to be patentable or to become intellectual property. Its purpose is to generate the knowledge base out of which the intellectual property of individual firms can emerge.

The problems of interest to industry in the precompetitive stage are now quite complex and require large-scale interdisciplinary efforts for progress to be made on a useful time scale. Usually, a team approach is indicated, with experts and facilities drawn from various disciplines.

The resources of most industrial laboratories are presently inadequate to mount such efforts in the number required by their businesses and on the scale required by the problems. Moreover, as implied by the discussion of Section IV there are powerful pressures to focus most of their resources on shorter-term activities with easily perceivable relations to the businesses and with less uncertainty with regard to the generation of intellectual property.

However, if the industrial laboratories do not participate actively in fundamental or basic research, leaving it all to universities and government laboratories, their companies will not recognize early on the opportunities for innovation presented by the results of that research, nor will they be able to influence it to advance in directions of interest and importance to their businesses.

The resolution of this dilemma is active leveraging of the resources the industrial laboratories do employ on precompetitive research by forming collaborations with individual researches and research groups at universities and government laboratories. To the extent that industrial researchers take the led in forming such collaborations, the directions of work will be of interest and the results ultimately of use to their companies.

In these collaborations, the university role is essential, but it has associated with it another dilemma. The present scale and vitality of university research has been achieved through the active participation of Ph.D. students and post-docs in that research. Through the less constrained growth of earlier years, the productive capacity for Ph.D.'s in physics of the university system has grown to the point where it exceeds the demand for new faculty and research physicists in a number of countries. Nevertheless, to sustain the university contribution to the above team approach to problems in physics of interest to industry, the scale of university research in physics should not decrease. Thus, the model of research in physics contemplated here would lead to the production of more Ph.D.'s than can be employed as physicists.

There is a resolution of this dilemma which has already been shown to work in specific cases. First, graduate students are not led from the beginning to expect research or university positions in physics as a matter of right after completing their degree. Second, the time for completion of the degree is reduced substantially below its typical present extent. Third, and most important, graduate study in physics should be regarded as advanced general education in the art of learning and of solving new problems, rather than training in a specific research area. Students should be informed that satisfying and rewarding jobs exist outside of physics for which they are well qualified and well regarded. It would help greatly in this endeavor if substantially more post-doc positions were available in industry in which the students could broaden their experience and better prepare themselves for a wider job search. One the industrial side, such post-docs could add substantially to the industrial contribution to the team effort and would not be as demanding of resources as permanent positions.

In summary, we have entered a new phase in the evolution of physics ... one in which its growth rate will fluctuate around that of the economy in the developed countries. Nevertheless, great opportunities abound for physics in industry, but adaptation of all research organizations to the changed circumstances must occur before they can be realized. Universities and government laboratories must form effective partnerships with industry, each playing its appropriate role, for society to realize benefit from those opportunities.

REFERENCES

[1] Price, D.J., "Quantitative Measures of the Development of Science". Archives Internationales D'histoire Des Sciences, 1951, Vol. 14, pp. 85-93.

[2] Price, D.J., "The Exponential Curve of Science", Discovery, 1956, No. 17, pp. 240-243.

[3] Price, D.J., panel discussion on "The Growth Of Scientific and Technical Information", Lecture Of J. Georges Anderla And Seminar Proceedings, National Science Foundation, January 1974, pp. 38-61.

[4] Price, D.J., "Science Since Babylon", 2nd Ed., Yale University Press, New Haven, 1975.

[5] Price, D.J., private communication

[6] Price, D.J., "Measuring the Size Of Science", Proceedings of the Israel Academy of Sciences And Humanities, 1969, Vol. 4, pp. 98-111.

[7] C. Holden, Science 257, 1710 (1992).

[8] J. Holusha, New York Times, September 5, 1993, p. F7.

[9] M. E. Porter, as quoted by S. Lohr, International Herald Tribune, September 3, 1992, p. 11.

[10] "Out of the Crystal Maze", L. Hoddeson, et al., Oxford University Press, New York, 1992.

[11] D. A. Bromley, Report to Congress on the State of Science, U.S. Gov't Printing Office, Washington, 1992.

[12] P. Hohenberg and W. Kohn, "Inhomogeneous Electron Gas", Phys. Rev. 136, B864-B871 (1964); W. Kohn and L. J. Sham, "Self-Consistent Equations Including Exchange and Correlation Effects", Phys. Rev. 140, A1133-A1138 (1965).

ACKNOWLEDGMENTS

Japan Society for the Promotion of Science, the Physical Society of Japan and the Japan Society of Applied Physics wish to express their appreciation to the following for their significant contributions in support of the XXI General Assembly of the International Union of Pure and Applied Physics.

Canon Inc.
Fujitsu Ltd.
Gakko Tosho Co., Ltd.
Hitachi, Ltd.
IBM Japan Ltd.
Kyoritsu Shuppan Co., Ltd.
Matsushita Electric Industrial Co., Ltd.
Matsushita Research Institute Tokyo, Inc.
Mitsubishi Electric Corporation
NEC Corporation
Nippon Steel Corporation
Nippon Telegraph and Telephone Corporation
Sanyo Electric Co., Ltd.
Sharp Corporation
Sony Corporation
Springer-Verlag Tokyo Inc.
Sumitomo Electric Industries, Ltd.
Sumitomo Heavy Industries, Ltd.
The Sederation of Electric Power Companies
Toshiba Corporation

NATIONAL ORGANIZING COMMITTEE
FOR
XXI IUPAP GENERAL ASSEMBLY

Hiroshi Takuma (Chairperson)
The University of Electro-Communications
Huzihiro Araki (Vice-Chairperson)
Kyoto University
Takuo Sugano (Vice-Chairperson)
Toyo University
Michiji Konuma (Secretary General)
Keio University

Jiro Arafune
The University of Tokyo
Akito Arima
Hosei University
Muneyuki Date
Osaka University
Hiroshi Ezawa
Gakushuin University
Tadao Fujii
Kogakuin University
Nobuhiro Go
Kyoto University
Kunio Hirata
Yamanashi University
Shin-ichi Hyodo
Meiji University
Yoshi-hiko Ichikawa
National Institute for Fusion Science
Hiroshi Kamimura
Science University of Tokyo
Tatsuyuki Kawakubo
Toingakuen Yokohama University
Kyozi Kawasaki
Kyushu University
Akira Kinbara
The University of Tokyo
Tetsuro Kobayashi
Tokyo Metropolitan University
Masao Koyanagi
Electrotechnical Laboratory
Eiichi Maruyama
Angstrom Technology Partnership
Shigeo Minami
Osaka Electro-Communication University

Toru Moriya
Science University of Tokyo
Sadao Nakajima
Tokai University
Isamu Shimizu
Tokyo Insitute of Technology
Hirosi Suzuki
The University of Electro-Communications
Jumpei Tsujiuchi
Chiba University
Akiyoshi Wada
Sagami Chemical Research Center
Hisatsune Watanabe
NEC Corporation
Yoshio Yamaguchi
Tokai University
Toshimitsu Yamazaki
The University of Tokyo

XXI IUPAP GENERAL ASSEMBLY

	Morning	Afternoon		Evening
Mon. Sept. 20	*Council Meeting*	*Council Meeting*		
Tue. Sept. 21	*Council Meeting*	*Council Meeting*	*Chair Persons and Council*	Reception Supper
	Registration			
Wed. Sept. 22	General Ass.	Academic Sessions		
Thu. Sept. 23	General Ass.	Academic Sessions		Banquet
Fri. Sept. 24	General Ass.	Excursion		Traditional Play
Sat. Sept. 25	General Ass.	*Council Meeting*		

Springer-Verlag
and the Environment

$\mathbf{W}$e at Springer-Verlag firmly believe that an international science publisher has a special obligation to the environment, and our corporate policies consistently reflect this conviction.

$\mathbf{W}$e also expect our business partners – paper mills, printers, packaging manufacturers, etc. – to commit themselves to using environmentally friendly materials and production processes.

$\mathbf{T}$he paper in this book is made from low- or no-chlorine pulp and is acid free, in conformance with international standards for paper permanency.

Lecture Notes in Physics

For information about Vols. 1–397
please contact your bookseller or Springer-Verlag

New Series m: Monographs